Secrets Of The Past...
Bridges To The Future

By

Annette Labovitz

With Carole Schaeffer

Central Agency for Jewish Education
Miami, Florida

Library of Congress Catalog Card Number: 1-317-473
ISBN 0-930-02900-3

For information address: Central Agency for Jewish Education
4200 Biscayne Blvd., Miami, FL 33137

Printed by James E. Fewquay at the
Greater Miami Jewish Federation

AUTHOR: ANNETTE LABOVITZ

EDITOR: JEROME L. HERSHON

COVER: BRACHA LAVEE (Jerusalem Artist)

Feltwork "Jerusalem"

ILLUSTRATION & MAPS: SHIRI ALYSSA

IN HONOR OF

My Parents

Yitzchak Elchanan Ben Eliezer and Rochel

and

Brientshe Bat Yehoshua Yisrael and Chava B'racha

WHO WERE MY FIRST TEACHERS

TABLE OF CONTENTS

.... CONTINUED ....

TABLE OF CONTENTS (CONTINUED...)

....CONTINUED....

TABLE OF CONTENTS (CONTINUED...)

ACKNOWLEDGEMENTS

ב"ה

Tammuz 5743
July 1983

I would like to gratefully acknowledge the following individuals without whose assistance this book could not have become a reality.

Rabbi Shlomo Carlebach who inspired me to tell stories,

Gene Greenzweig, Executive Director of the Central Agency for Jewish Education, whose foresight and vision provided guidance,

Abraham J. Gittelson, Assistant Director of CAJE, for his encouragement,

Rabbi Jay J. Schachter who checked the manuscript for historical accuracy,

Reb Moshe who checked the correctness of the quoted Biblical, Talmudic and Midrashic passages, often locating the most hidden and mysterious sources,

Mr. Jerome L. Hershon who painstakingly edited the manuscript,

Judge Theodore Nelson for handling all the legal aspects associated with publishing this book,

Ruth Eisenberg, publications director of CAJE, who created a book from a manuscript,

My husband and children for their patience and cooperation.

Annette Labovitz

Preface

This is a book of stories about our glorious Jewish past. There are many versions of the stories which appear in this book. I have tried, to the best of my ability, to be faithful to the facts of the stories as I know them. I have added dialogue and dramatization to heighten the excitement and enhance the theme.

Stories touch the emotions of young and old alike. They describe the unique experience of the Jewish people during the past two thousand years; they integrate the panorama of "Western" civilization with the Jewish presence; they bring understanding to being Jewish in a changing world. Indeed, stories are "Secrets of the Past, Bridges to the Future".

I will begin this book by allowing the "Rebbe of Rizhin" to explain why stories are such an important part of the tradition of our people.

Annette Labovitz

THE RIZHINER REBBE EXPLAINS THE IMPORTANCE OF STORIES[1]

Rebbe Yisrael Rizhiner, the Rebbe of Rizhin, scion of the royal house of King David, resembling his great namesake, Rebbe Yisrael Baal Shem Tov, the founder of the Chassidic movement, lived regally and conducted his Chassidic court majestically. He loved to tell stories. He also loved to listen to stories. Rich and poor, learned and unlearned, approached him for comfort, guidance and advice.

The Rebbe of Rizhin had a special appointment secretary who arranged his schedule. One day, two men entered the anteroom of the Rebbe's chamber simultaneously and requested an appointment with the Rebbe.

One man explained to the appointment secretary that he had just completed writing a commentary on Jewish law. He wanted to show his manuscript to the Rebbe. He hoped the Rebbe would honor him by writing the preface to the book.

The second man explained to the appointment secretary that he had just completed writing a book of stories. He also wanted the Rebbe to write the preface to his book.

The secretary was perplexed. Who would the Rebbe see first? In his mind, he was certain that the Rebbe would choose to see the author of the commentary on Jewish law first. However, he decided to ask the Rebbe before he scheduled the appointments.

The appointment secretary was very surprised when Rebbe Yisrael asked him to schedule the story teller for the first interview.

> "You see," explained the Rebbe, "in meeting with these two men, I wish to follow the order of our Torah. The five books of the Torah begin with stories: B'rayshit (Genesis) is the story of Creation and the origin of the Jewish people, and Sh'mot (Exodus) is the story of enslavement, freedom, and Revelation on Mt. Sinai. The laws begin after the story of the Revelation and continue through the third book of the Torah, Vayikra (Leviticus) which teaches the laws between man and G-d, and man and man. Therefore, without the stories, there would be no foundation for understanding the laws we observe."

The appointment secretary understood Rebbe Yisrael's reasoning and scheduled the storyteller for the first interview.

1 This story was adapted.

THE STUDY OF TORAH
תלמוד תורה

CHAPTER ONE

"This Torah shall not depart from your mouth,
and you shall meditate upon it day and night."[1]

"For it is our life and the length of our days."[2]

The Torah has guided the Jewish people throughout their history on the path of holiness, so that they might become a "kingdom of priests and a holy nation".[3]

The Hebrew root of the word Torah, 'yaroh', means to teach and instruct, and it is Torah which has taught and instructed the Jewish people. The Torah, that precious gift which Moshe Rabaynu (Moses our teacher) brought down from Sinai, has been the source of holiness in Jewish life and the secret strength of both physical and spiritual religious survival. It is "our life and the length of our days".[4]

From the time of the Revelation at Sinai, the Torah has been transmitted through an unbroken chain of tradition from generation to generation, linking the past to the present and the present to the future. Outstanding Torah teachers have led the Jewish people through the cross currents of Jewish history, guiding, advising and instructing, in order that the chain of tradition be perpetuated.

These Torah teachers were imbued, not only with the love of Torah, but with the knowledge that Torah guaranteed the eternity of the Jewish people. Accordingly, they left ethical wills delineating the importance of Torah study. Moses instructed the Levites "to teach G-d's laws to the House of Jacob and G-d's teaching to the people of Israel"[5].

King David pleaded with G-d "to teach me Your Ways that I may walk in Your Truth"[6], and Rabbi Chiya directed that "the study of Torah should be the major occupation of the Jewish people"[7]. Thus, the tradition of studying the Torah became one of the major vocations of the Jewish people.

Throughout the long span of Jewish history, Jews succeeded in studying Torah and transmitting it to their children. Even in the face of religious persecution, oppressive restrictions, and inhuman treatment, the study of Torah was never abandoned; and thus preserved Jewish life.

The study of Torah strengthened the weary, comforted the wanderers and brought hope to the downtrodden, whose faith kept the Jewish people alive.

> "A flame burns in the fireplace,
> The room warms up,
> As the teacher drills the children in the aleph-bayt.
> Remember dear children,
> What you are learning here, repeat it again and again,
> Kametz aleph is pronounced AW.
> When you grow older you will understand that this alphabet
> Contains the tears and the weeping of our people.
> When you grow weary and burdened with exile,
> You will find comfort and strength within this Jewish alphabet."[8]

The stories in this chapter are about great Jewish sages who prescribed the study of Torah as an antidote to persecution and assimilation because they believed that Torah is a "tree of life to them who cling to it".[9]

1 Joshua 1:8
2 Deuteronomy 30:20, Talmud Bavli Berachot 61 b
3 Exodus 19:6
4 Deuteronomy 30:20
5 Deuteronomy 33:10
6 Psalms 86:11
7 Talmud Bavli Berachot 16 b
8 Mark M. Warshawsky, Oif en Pripetchok, At the Fireplace. Jewish Songs. Metro Music Co. New York. (The songs of Mark M. Warshawsky were originally published in eastern Europe. By the turn of the twentieth century, they became household favorites in Jewish homes. They appear in many collections of Jewish music.)
9 Proverbs 3:18

ELIYAHU INSPIRES A FISHERMAN[1]

According to a Midrash, G-d consulted with Eliyahu before He decided to create man. The Almighty asked Eliyahu:

"Shall I create Man?"

Eliyahu answered:

"If it appears good in Your sight, it is certainly good in mine, and if You wish it, I will go down into the world and serve man by teaching him Torah."[2]

Since that time, even to this very day, Eliyahu roams the world as an Heavenly emissary, teaching Torah. This story is but one tale of the popular folklore that grew up around the man who taught the righteous and inspired the simple.

One day Eliyahu was walking along the banks of a river when he saw a man standing on a nearby wharf. Eliyahu approached the man and asked:

"How do you earn your livelihood?"

The man answered:

"I am a fisherman."

Eliyahu further asked:

"What else do you do in addition to fishing?"

The fisherman bashfully answered:

"Nothing."

Eliyahu then asked the fisherman if he had ever studied the Torah. The fisherman began to cry, and said:

"I have never learned Torah, nor do I think I have the capacity to understand the depth of Torah."

Eliyahu then asked:

"How did you become the greatest fisherman on this wharf?"

The fisherman replied:

"I spend my days and nights, every waking moment, every ounce of my strength fishing, because my life depends on it."

Now Eliyahu began to cry, and said:

"Oh, fisherman, if only you knew that there is more to life than fishing...that the real joy of life depends on studying the mysteries of G-d's Torah...If only you spent one tenth of the time on studying G-d's law as you do on fishing, you would be not only the greatest fisherman on this wharf but the greatest scholar that walked on the face of the earth."

Whereupon, Eliyahu left the inspired fisherman and continued on his journey to teach Torah to whomever else wanted to learn.

The fisherman followed Eliyahu's advice and decided to devote a specific amount of time each day to the study of Torah.

1 One of the sources of legends about Eliyahu Hanavi is the Midrashic work, Tanna Devay Eliyahu, which is divided into two sections: "Sayder Eliyahu Raba" and "Sayder Eliyahu Zuta". This story was adapted from "Sayder Eliyahu Zuta".

2 Yalkut Reuveni, B'Rayshit. ed. Amst. p. 98

HILLEL BECOMES A SCHOLAR[1]

Throughout Jewish history, adversaries who understood the secret of Jewish survival attempted to restrict the study of Torah and the observance of certain "mitzvot" (commandments) which were the essence of a Jewish life. Their goal was to eradicate Judaism.

The Greeks, around the fourth century B.C.E., were among the first to attempt to achieve this goal by decree, but they were unsuccessful.

The greatest threat to Jewish survival came between the end of the first century B.C.E. and the beginning of the fifth century C.E., from the Romans who conquered and ruled Eretz Yisrael.

The Romans realized that the only way they could subdue the Jewish people was to destroy Judaism. They thought they could best achieve this goal by prohibiting the study of Torah, eliminating the educated Torah observing Jew, and destroying the Jewish will to resist the alien values of Roman culture. The Romans failed.

The Rabbis transmitted Torah through the "Mishna" (the exposition of Torah law), directed the building of synagogues in the villages of Eretz Yisrael, oversaw the writing of the liturgy, and set up schools for Torah instruction.

Because of the Jews' stubborn refusal to give up their precious possession of Torah, they outlasted the mighty Roman Empire.

When Hillel Hazaken, the Elder, arrived in Eretz Yisrael from Babylonia about one hundred years prior to the Roman destruction of Jerusalem in 70 C.E., he already found Jewish life in turmoil because of the Roman conquest. He was the scion of the royal house of King David, and although he was born in Babylonia, he decided to emigrate to Eretz Yisrael so that he might have the opportunity to study Torah with the great teachers of the time.

When Hillel reached Eretz Yisrael, he was very poor. He had very little money to sustain himself. He found a job as a woodcutter for which he was paid a half a dinar a day, a very meager sum. He used half the money for his personal sustenance and half to pay the daily price of admission to the study hall of Shmaya and Avtalyon, two outstanding Torah teachers with whom he wanted to study.

One day, Hillel was not able to earn any money, and consequently, he could not pay the price of admission to the gate keeper at the study hall. He did not want the day to go by without studying some Torah, so he climbed up to the roof in order to hear through the skylight what little he could of the lessons being taught by Shmaya and Avtalyon.

It was the cold winter month of Tayvayt, and snow began falling. Hillel, lying on the skylight, with his ear pressed on it, was so absorbed in trying to understand the lesson that he was oblivious to the falling snow. The cold had no effect on him, but after a while, the deep concentration caused him to fall asleep.

The skylight was the main source of light in the little study hall. When Shmaya and Avtalyon arrived the following day, they could not understand why the usually lit room was dark. They walked outside to check if the day was cloudy, and when they looked up, they noticed the figure of a man lying on the roof.

Shmaya and Avtalyon climbed up to the roof to investigate further. They found the almost frozen Hillel, covered with three cubits of snow, sleeping with his ear flat on the roof as he had been straining to listen to the words of Torah. Shmaya and Avtalyon carried the frozen Hillel down from the roof and brought him into the study hall. They revived him and gave him warm food to eat. They understood how desperately Hillel wanted to learn Torah, so they granted him a full scholarship to the study hall from that time on.

Hillel became a great scholar, and subsequently the president of the Sanhedrin, the Jewish Supreme Court. He exercised authority not only over religious matters, but also over ethical, political and economic affairs as well. He was an inspiring model, both patient and humble, constantly bringing people closer to the Torah way of life.

1 Talmud Bavli Yoma 35 b

RABBAN YOCHANAN BEN ZAKKAI SAVES YAVNEH[1]

The Roman seige of Jerusalem in the years 67-70 C.E. precipitated a crisis in Jewish life. Many Jewish people felt that the physical destruction of Jerusalem was inevitable and it would mark the annihilation of the nation. However, some people, led by Rabban Yochanan ben Zakkai, felt that the survival of Judaism depended on the salvation of our spiritual heritage, the Torah. Rabban Yochanan ben Zakkai improvised a plan whereby Torah might survive.

His plan was to escape from the besieged city of Jerusalem in order to approach the Roman general Vespasian and plead with him to spare the village of Yavneh and its sages. He put his plan into effect by feigning illness and pretending to be on his death bed. His nephew, Abba Sikra, executed the plan. He announced Rabban Yochanan's illness and requested that the great Rabbi's students visit him frequently. Then he obtained a piece of decayed meat which had the odor of a dead body. With these necessary preparations completed, Abba Sikra announced his uncle's death.

Eliezer and Joshua, two trusted disciples of Rabban Yochanan, proceeded to arrange for the funeral of their teacher. They placed Rabban Yochanan and the decayed meat in a casket. Abba Sikra, Eliezer and Joshua carried the casket to the gates of the city. The burial grounds were outside the wall of Jerusalem, and Rabban Yochanan's disciples insisted that it was imperative to bury their teacher outside the wall. Rabban Yochanan had instructed them that only they should carry the casket because the Roman guards might lift the casket and they would suspect the weight of the casket to be too light for a dead body. When they reached the gates, the guards insisted on stabbing the body to ascertain death, because no one was allowed to leave Jerusalem alive. Abba Sikra, Eliezer and Joshua convinced the guards that it is prohibited by Jewish law to mutilate a body, and the guards allowed them to carry the casket through the gates.

Immediately upon his escape, Rabban Yochanan made his way to the headquarters of the Roman General Vespasian. He greeted the general by saying:

"Peace unto you King, peace unto you."

Vespasian replied:

"You deserve death twice. First, you address me as King, and I am not a King and second, if I am King, why haven't you come to honor me and pay your respects till now?"

Rabban Yochanan answered:

"In answer to your first question, I claim that you are the King, and in answer to your second question, I must tell you that there are many among us who tried to prevent me from approaching you."

While they were talking, a messenger arrived from Rome and relayed the message that the Roman senate had indeed chosen Vespasian as Emperor, confirming Rabban Yochanan's statement.

Vespasian turned to Rabban Yochanan ben Zakkai and said:

> "Since you are so wise, ask me some favor, and I shall try to grant it."

Rabban Yochanan answered:

> "Grant me three things. Save the village of Yavneh with its sages; spare the lives of Rabban Gamliel and his family who are descendants of the noble house of King David; and send a doctor to cure the illness of Rabbi Zadok who fasted a very long time in order that Jerusalem might not be destroyed."

Vespasian granted Rabban Yochanan ben Zakkai's requests and returned to Rome. Yavneh became one of the great centers of Torah learning.

1 Talmud Bavli Gitin 56 a

THE FARMER WHO BECAME A GREAT SCHOLAR[1]

Many Talmudic stories describe the struggle of our ancestors to survive Roman rule by turning against wealth and politics in order to delve into the mysteries of Torah. This story is about Rabbi Eliezer ben Hyrcanus who was one of the five most prominent disciples of Rabban Yochanan ben Zakkai. He devoted his talents to the study of Torah and the leadership of the Jewish people.

Rabbi Eliezer ben Hyrcanus was not always a scholar. In fact, he did not begin to study Torah until he was twenty-eight years old. Until then, he earned his livelihood as a farmer, helping his father and brother plow the fields and reap the harvests.

One day as he was plowing, he noticed that his brother was working on level land and he was working on the hilly side of a mountain. He sat down and cried. His father asked him:

> "Why are you crying? Is it because you are plowing on the hill and your brother is plowing level land? Come, from now on, you can also plow on level land."

Days later, Eliezer was plowing on level land, but still crying. His father wondered at the cause of his son's unhappiness and asked Eliezer why he was crying this time. Eliezer answered:

> "I want to go to Jerusalem to study Torah. If I have the strength to plow fields, I certainly have the strength to study Torah."

His father rejected Eliezer's plea. He said:

> "Now is not the time to study. You are old enough to get married. When you will have children, you will take them to the study hall and they will learn Torah."

Eliezer was not satisfied. He was determined to learn Torah in Jerusalem in the school of Rabban Yochanan ben Zakkai.

His father was very angry and threatened to cut off Eliezer's inheritance. He thought that Eliezer would forget his foolish ideas if he knew how much his father disapproved of his plans.

Eliezer had other ideas. He finished his day's plowing and set out for Jerusalem. He was so determined to study in Jerusalem that it did not matter to him that he had no food or money. When he was hungry, he ate handfuls of dirt from a sack which he carried with him in order to still the hunger pangs in his stomach.

Eventually, he reached Jerusalem. He enrolled in the school of Rabban Yochanan ben Zakkai and sat unobtrusively at the edge of the study hall. He listened intently to the lesson, but didn't understand a word. Eliezer began to sob and Rabban Yochanan ben Zakkai noticed it. After the lesson, Rabban Yochanan ben Zakkai walked over to the young man and asked why he was sobbing.

Eliezer answered:

> "I desperately want to learn Torah, but I don't understand the lessons."

Rabban Yochanan ben Zakkai realized that Eliezer had never learned the Hebrew alphabet, the recitation of the Sh'ma, or the grace after meals. He began to teach Eliezer these basic fundamentals of Judaism on the spot.

Eliezer learned with great diligence. Soon, he was able to read prayers and understand basic concepts. He became so absorbed in his studies that he forgot to eat for eight days.

One day while Rabban Yochanan ben Zakkai was teaching Eliezer, he detected a foul odor being emitted from Eliezer's mouth. Rabban Yochanan realized that his new student had not eaten food for a lengthy period. He asked Eliezer if he had eaten that day, but the young man was too embarrassed to admit the truth.

Rabban Yochanan ben Zakkai sent two students to the boarding house where Eliezer slept to ask the landlady if she was providing food for Eliezer. She said that she assumed he was eating with Rabban Yochanan ben Zakkai. Then she pointed to a sack and said:

> "At times, I saw him nibbling from this sack."

The students opened the sack, and realizing that Eliezer was eating dirt, raced to inform Rabban Yochanan ben Zakkai.

Rabban Yochanan ben Zakkai said to Eliezer:

> "Until now a foul odor was emitted from your mouth because you hadn't eaten for many days. Now I bless you that you will become a great scholar. Only goodness will be forthcoming from your mouth. From now on, you shall be a guest at my table."

Eliezer remained with Rabban Yochanan ben Zakkai for many years. Rabban Yochanan ben Zakkai's blessing was fulfilled.

One day, Rabban Yochanan ben Zakkai assembled the scholars of Jerusalem to discuss words of Torah. He noticed a stranger in the crowd who had not been invited and recognized him to be the father of Eliezer. Before the stranger said a word, Rabban Yochanan instructed a student to seat him in a place of honor.

Then Rabban Yochanan ben Zakkai asked Eliezer to deliver a Torah discourse. His words brought joy to the crowd, and when Eliezer finished, Rabban Yochanan rose and kissed him on the head.

Then Eliezer's father rose and with great joy, he said:

> "I came to Jerusalem to complain to this honored assembly that my son left me in my old age. I am overjoyed that he rejected my advice and came to Jerusalem to study. I am proud that he became a great scholar and I now bequeath him all of my fortune."

Rabbi Eliezer ben Hyrcanus said:

> "Father, if I wanted wealth, I could have beseeched the Almighty. I did not come to Jerusalem for wealth or honor, only to study Torah. Please bequeath your wealth to my brother."

Rabbi Eliezer ben Hyrcanus remained in Jerusalem and became a prominent leader among the Jewish people. Rabban Yochanan ben Zakkai said of his disciple:

> "If all the sages of Israel were on one side of a scale, and Rabbi Eliezer ben Hyrcanus was on the other side, he would outweigh them all."[2]

1 B'Rayshit Raba 42:1
2 Avot 2:8

RASHI, GREATEST OF BIBLICAL COMMENTATORS[1]

During the early part of the Middle Ages, the Jews were welcome residents in the towns and villages of the Germanic lands (France and Germany). The most popular way for them to earn a living was as merchants, and in some towns and villages, special sections were assigned to them for dwellings so that the progress of their economic activities would be unhindered by the peasants. A network of trade routes developed, not only throughout the area of the Germanic lands but also throughout the lands bordering the eastern shores of the Mediterranean Sea. The known world was linked by trade routes.

In the countryside, feudal masters patronized Jewish traders for the mutual benefit of both. If a feudal master was not treating 'his Jews' with honor, all the Jews in that area left and vowed never to return. They could usually find an area where they would be guaranteed better treatment.

Highway robbers and murderers knew it was of no advantage to harm Jews, because the Jewish community pressured the feudal master until the culprit was punished. Since the feudal master collected the fines imposed on the culprit, mutual cooperation existed. Jews had a pact never to buy or sell property stolen from another Jew.

Each Jewish community had its leaders who legislated for the mutual benefit of all inhabitants, and who also served as liaison between the community and the feudal master. The Jewish leaders collected taxes due to the feudal masters for the privilege of living in their domain, controlled business practices, appointed judges and enacted laws to strengthen religious observance.

In addition to a measure of economic success, the Jews were free to delve into the study of Torah. Great Yeshivot developed in the communities that were situated along the Rhine River. The Yeshivot of Speyer, Worms, and Mainz were famous throughout the Germanic lands, and students flocked to these Yeshivot to study with its scholars.

Many scholars and Torah commentators lived during this period in Jewish history. The most outstanding was Rabbi Shlomo ben Yitzchak, Rashi. His Torah and Talmudic commentaries formed a lasting contribution to the transmission of Torah from generation to generation. Rabbi Yitzchak HaLayvi, one of Rashi's teachers proclaimed:

> "This generation is not orphaned (from Torah)
> because you are representative of it,
> May there be more like you in Israel."[2]

Many legends have been told about the birth and life of Rashi, who was born in Troyes in the province of Champagne, France, in the middle of the 11th century. His father, Rabbi Yitzchak, was a wealthy merchant and the head of a distinguished family.

Rabbi Yitzchak owned a precious diamond. The priest of the city was anxious to buy the diamond to decorate the eye of a statue that stood in the vestry of the church. The priest offered Rabbi Yitzchak a great deal of money for the diamond, but he refused to sell it because he knew it would be used for idolatrous purposes. Finally, the priest lured Rabbi Yitzchak onto a boat, and when they were far from shore, he attempted to coerce Rabbi Yitzchak to sell the diamond. In desperation, Rabbi Yitzchak threw it into the sea.

When Rabbi Yitzchak returned to shore, he heard a Heavenly Voice proclaim:

> "Since you did not permit the diamond to be used to decorate the eye of the statue, you will be rewarded with a more precious diamond that will be a sparkling source of light to the Jewish people."

The news of the Heavenly Voice quickly spread throughout the city of Troyes. When certain people in the Christian community realized that Rabbi Yitzchak's wife was pregnant, they attempted to kill her, to abort the proclamation of the Heavenly Voice.

Rabbi Yitzchak and his wife realized they were in great danger, so they decided to leave Troyes temporarily. Rabbi Yitzchak arranged for his wife to leave the city immediately, and he planned to follow her as soon as he could settle certain business affairs. He hired a carriage to take her to Mainz, Germany, the city where her brother Rabbi Shimon Hazaken lived.

On the way, the carriage stopped in Worms. Rabbi Yitzchak's wife realized that she was beginning to have labor pains, so she left the carriage and began to wander through the city in search of someone in the Jewish community who would help her. She walked through streets and narrow alleys. Suddenly, she realized that she was standing in the way of a rushing coach driven by a feudal lord. She turned to face the wall, fearing that the coach would crush her. Just at that moment, she felt the wall heave and the coach passed through without hurting her. She gave birth to her baby son at that place in Worms where the wall moved.

The Jewish community heard of the wondrous deed which G-d wrought. They made the mother and child comfortable and sent a letter to Rabbi Yitzchak to inform him of the birth of his son. He arrived in time for the circumcision and he named his son Shlomo in memory of his father. After a while, the family returned to Troyes.

The boy had an innate thirst for studying Torah. His family encouraged his learning, and soon, Shlomo had a reputation of being an "ilui" (an intellectual genius). He was recognized for his ability to verbalize the lengthy and complicated commentaries of the Torah that had accumulated over the centuries into simple and concise explanations. When his knowledge surpassed that of the scholars in his city, his family permitted him to travel to Torah academies in other cities. While he traveled, he meticulously recorded the "halachic" (legal) opinions of the scholars of the Torah academies he visited.

Shlomo returned to Troyes at about the age of twenty-five and established his own academy. It did not take long for him to be recognized as a Talmudic authority. He came to be known as Rashi, the acronym of his name, Rabbi Shlomo ben Yitzchak.

In Troyes, Rashi sifted the massive accumulation of notes which he gathered during the years he travelled, and he wrote his concise explanation of the Torah, the Prophetic Writings and the Talmud. Rashi's writings became classics, and his fame spread far beyond his native land. A disciple commented:

> "Rashi was the father, the teacher, and the revealer of hidden secrets. Without him, the Talmud would remain a sealed book."[3]

1 Chumash Bayt Yehuda Hashalem: B'Rayshit, Vol. I. "Toldot Rashi". Hotzaat Orah. New York. 5714 (1954). This book is a collection of Yiddish commentaries on the Torah. There is no single author.

2 Talmud Bavli Chagiga 3 b. This quote was originally said about Rabbi Eleazar ben Azarya, familiar as one of the five Rabbis who sat at the Passover Sayder table in Benay Brak retelling the story of the Exodus in the face of the Roman prohibition to refrain from such activities. The other Rabbis were Rabbi Eliezer, Rabbi Yehoshua, Rabbi Akiva and Rabbi Tarfon.

3 Rabbi Bezalel Ashkenazi. J.D. Eisenstein. Otzar Yisrael. Vol. X. Shilo Publishing Co. Jerusalem.

EDEL'S SHMUEL[1]

During medieval times, the rulers of Poland - Lithuania gave precedence to the economic development of their country, so they sought immigrants from western Europe to help them build their land. King Casimir the Great invited Jewish refugees who fled western Europe to escape being blamed for the Black Death Plague and the Blood Libels to settle in his lands. Later kings welcomed the Jews who fled from the Spanish Inquisition in search of a safe place to live.

The Jewish immigrants were granted equal citizenship in Poland, and they lived a generally secure and content life. Polish Jewish life was so comfortable during the fifteenth and sixteenth centuries that Rabbi Moses ben Israel Isserles (the Rama), leader of the Cracow, Poland Jewish community and a great codifier of Jewish law stated that:

> "It is preferable to live on dry bread and peace in Poland than to live in luxury in lands which harbor dangers for Jews."[2]

To show his feelings about the security which Jews felt in Poland, he devised a play on the word Poland from its Hebrew equivalent, "Poh-lin" (here we will lodge).

The tranquility of Jewish life in Poland led to the establishment of great "Yeshivot" (great schools of Talmudic learning). These "Yeshivot" were originally established by Jewish scholars who fled from western Europe. Eventually, the Jewish community in Poland produced its own native scholars and Rabbinical leaders.

Torah study received the highest priority in Jewish life. It was generously supported by the community. In the study halls of the "Yeshivot", there was no difference between rich and poor. Recognition was accorded for the degree of achievement in Talmudic scholarship.

Great Torah scholars illuminated Polish Jewish life, among them prominent Talmudic commentators. One of the indispensible Talmudic commentaries of this period might not have been written had it not been for the foresight and determination of the natural parents of Rabbi Shmuel Eliezer ben Yehuda Edels who were childless for many years after their marriage.

The story begins in the city of Ostrog, Poland where a new synagogue was being dedicated. The city people gathered with the people from the surrounding villages to share the joy of this momentous occasion. Everyone was crowded into the new synagogue building. As the ceremony proceeded, a wealthy man from one of the nearby villages stepped forward from the crowd to the pulpit in the center of the synagogue and offered to donate a tremendous sum of money for the upkeep of the synagogue. Many other people were inspired and followed his example by promising additional sums of money.

The Rabbi turned to the wealthy man and whispered to him that he wished to see him privately after the ceremonies.

During their visit, the Rabbi thanked the wealthy man for his tremendous donation and proceeded to inquire if there was anything he could do for him.

"You can ask me for anything," said the Rabbi, "and I will pray for the fulfillment of your desires."

The wealthy man spoke thoughtfully:

"Yes, as a matter of fact, you can do something for me. You can pray that my wife and I be blessed with a child to carry on our family name."

The Rabbi's face darkened: he never expected that the wealthy man would ask for something so difficult to fulfill. He said:

"Please come back after three days have passed. At that time, I will advise you what to do."

On the third day, the wealthy man returned to the Rabbi's house. The Rabbi himself opened the door and escorted the wealthy man to his study. The Rabbi said:

"It is possible for your desire to be fulfilled under three conditions."

"What are the conditions?" asked the wealthy man.

The Rabbi answered:

"First, you will lose your wealth; second, when your wife becomes pregnant, you will die; and third, shortly after the child is born, your wife will also die."

The wealthy man was astounded. He turned to the Rabbi and said:

"Acceptance of these conditions does not depend only on me. Since it concerns my wife, let me consult with her."

The Rabbi and the wealthy man took leave of each other. The wealthy man returned home with mixed emotions and revealed to his wife the difficult conditions they would have to accept in order to be blessed with a son. The wife unhesitatingly agreed to accept the conditions, so the wealthy man returned to the Rabbi and informed him of their decision. The Rabbi blessed the wealthy man and said:

"The child that will be born to you will be a source of great light to the Jewish people."

Almost as soon as he returned home from his visit to the Rabbi, the wealthy man's business began to fail. It did not take long for him to lose his great wealth. After a few months, his wife became pregnant, and the former wealthy man passed away, just as the Rabbi had foretold.

The widow remained alone in the village. She was comforted only by the Rabbi's promise of the greatness her yet unborn child would achieve. As the time for the baby's birth approached, she decided to go to the city of Posen. She thought that there might be some compassionate people in the city who would assist her and help her care for the child.

She was embarrassed to enter the city during the day lest anyone recognize her, so she went in the blackness of night. She saw a faint light coming from a house on the edge of the city and recognized it to be the baker's house. She recalled that the baker had once been the recipient of her husband's generosity, so she knocked on the door. Because the baker was hard at work preparing rolls for his customers, his wife opened the door. She looked at the pregnant lady with the luxurious clothing that had become tattered and worn, and immediately had compassion for her. Recognizing the pregnant woman to be Rachel, the widow of the wealthy man, she invited her into the house and made her as comfortable as possible.

The baker and his wife cared for Rachel as if she were their own daughter. A few days later, a baby boy was born. The baker and his wife arranged for the circumcision. The boy was named Shmuel Eliezer.

The baker and his wife helped care for the infant and they became very attached to him. They remembered the many favors the boy's father had done for them, and they hoped that this child had inherited his father's kindness and wisdom. They had a daughter named Devorah and they decided to contract a match between themselves and Shmuel's mother for the children to be married when they reached the proper age.

Realizing that the third condition that the Rabbi foretold would soon be fulfilled, Rachel agreed to the engagement contract between herself and the baker and his wife on the condition that they would accept guardianship of her son upon her death.

When Shmuel was three months old, his mother passed away. He was raised by the baker and his wife. The name of the baker's wife was Edels and the boy became known as "Edel's Shmuel". He was intellectually more advanced than other boys his age. When he reached marriageable age, the baker and his wife revealed to Shmuel that they were his guardians, not his parents, and that his mother had contracted his marriage to their daughter before her death.

He married Devorah, the baker's daughter. His mother-in-law helped him establish a "Yeshiva" in Posen and he attracted many students. As his scholastic ability was recognized, it became obvious that the blessing of the Rabbi of Ostrog was fulfilled. Rabbi Shmuel Eliezer ben Yehuda Edels became known as the Maharsha, one of the great Talmudic luminaries of that age. Ultimately, he moved to Ostrog where he became head of the largest "Yeshiva" in Poland.

This passage was inscribed over the doorpost of his home:

> "The stranger shall not lodge in the street.
> I shall open my door to the traveller."[3]

1 M.S. Kasher and Y. Belkrovitz. Anshay Mofayt. M.S.D. Publishing. Jerusalem. 1976. This story was adapted.

2 Moses ben Israel Isserles. The Responsa of the Rama. Responsa 73. Cracow. 1640.

3 Job 31:32

THE OPEN PAGE OF TALMUD BAVLI, TRACTATE BECHOROT, PAGE 19 b[1]

Since the time that the Eternal labeled His people "am keshay oref" (a stiff-necked people)[2], we have stubbornly persisted in our fight for Jewish survival.

From our experiences throughout Jewish history, we have come to learn that Jewish survival is inextricably bound with the study and promulgation of Torah. Indeed, "Kudsha berich Hu, oraita V'Yisroel chad Hu" (G-d, the Torah, and Israel are One)[3].

Rabbi Akiva recognized this dictum two thousand years ago when he presented this parable to Papus ben Yehuda who questioned his tenacity to teach Torah in the face of the Roman decree forbidding it.[4]

> "A fox was walking along the shores of a river and noticed that the fish were scurrying from shore to shore, as if in great fear.
>
> "The fox said to the fish:
> 'Why are you scurrying from shore to shore?'
>
> "The fish answered:
> 'We are fleeing from the nets of the fisherman.'
>
> "The fox said:
> 'Why don't you float to the dry land and we will all live together in peace and harmony?'
>
> "Whereupon the fish answered:
> 'You, who are called the cleverest of animals, are indeed very foolish. We are scurrying from shore to shore trying to find safety in our natural environment. Don't you think we will encounter many times more life threatening dangers outside of our natural habitats?'"

The application of this philosophy, i.e. the studying and promulgation of Torah in whatever circumstances we may find ourselves, has guided us through each phase of our history. So it was during the horrors of the Holocaust when an Austrian born drifter named Adolph Hitler led the Third Reich in decimating European Jewry. The Holocaust brought to an end the creative religious and secular life of a Jewish community that had planted its roots upon European soil some two thousand years before.

Nevertheless, during all the years of the Holocaust, Torah was studied in hidden cellars, attics, and secluded rooms in the ghettos of Kovno, Lodz, Vilna and Warsaw, and in secret cells and concealed bunkers in the concentration camps at Auschwitz, Bergen Belsen, Dachau, Maidenek and Theresienstadt. The young people who risked their lives to perpetuate Torah did so because the study of Torah gave them the will to survive.

This story is about Rabbi Avraham Weinberg, who persisted to study Torah until the moment the Nazi murderers rounded him up together with his students and transported them to Treblinka. They were rounded up so quickly that they did not have the opportunity to close their books of the Talmud. In Yeshivot in the United States and Eretz Yisrael, a new generation of students are perpetuating the study of Torah. Some of them begin their Talmudic learning on the page where Rabbi Weinberg left off.

Rabbi Weinberg did not deviate from his daily routine of learning, even though he was "imprisoned" in the Warsaw ghetto which was under the control of the Nazi murderers. To the contrary, he continued to dedicate himself even more to the study of Torah than he had before. He immersed himself in learning day and night. Those people who knew of his dedication to Torah said that each time he outmaneuvered the Nazi decree to stop studying, the greater was his personal sense of triumph.

His lessons in the secret study hall continued on a daily basis. Groups of students, some very young, some well versed, gathered to listen to his lessons. He sat surrounded by eager students opposite his open copy of the Talmud which had been hidden when volumes of holy books were confiscated by the Nazis in order to gather them in a museum to exhibit remembrances of Jewish culture. It was evident that the joy of learning Torah was responsible for making him forget hunger, physical deprivation and the infliction of mental anguish.

Rabbi Avraham Weinberg accepted his personal suffering with equanimity, but through the power of his prayers he tried to alleviate the anguish of other Jews whose pain seared his heart.

The number of Rabbi Weinberg's students increased from day to day as newcomers flocked to his secret study hall. They arrived from other areas of Poland that had gradually succumbed to Nazi domination, and were herded into the Warsaw ghetto along with over a half million other Jews. These students were physically hungry for bread but they were spiritually thirsty for Torah.

As the war progressed, the situation in the Warsaw ghetto worsened. The Nazi murderers began transporting the inhabitants of the ghetto to the death camps. The few people that were taken to labor camps were forced to toil for German industry, and some survived; but those who did not receive work cards were excluded from the privilege of living. According to the Nazi murderers, they were declared "illegals".

Many people tried desperately to obtain work cards, but Rabbi Weinberg's students preferred to continue learning Torah rather than take the time out to obtain them.

The time for the final liquidation of the ghetto was rapidly approaching. The murderers with their superior armaments surrounded whole sections of the ghetto, drove the inhabitants out of the hiding places, and marched them to the trains which transported them to

Treblinka. Rabbi Weinberg refused to move. He showed no interest in attempting to save himself. He would not leave his students without a teacher. He continued to teach those who escaped the round up. Their numbers diminished daily.

This way he sat and learned. One day, the murderers detected his secret study hall and herded Rabbi Weinberg and his remaining students into the trains that were headed for Treblinka.

When Rabbi Weinberg walked out of the study hall, he left his volume of the Talmud Bavli open to Tractate Bechorot 19 b, for posterity.[5]

1 Mordechai Eliav. Ani Maamin. Mosad HaRav Kook. Jerusalem. 1969. This story was adapted from testimonies of the survivors of the Holocaust.
2 Exodus 32:9
3 Zohar, Acharay Mot, Chapter 73
4 Talmud Bavli Berachot 61 b
5 Talmud Bavli Bechorot 19 b. On this page of this tractate there is a discussion betweem Rabbi Acha, Rabina and Abaye about the qualifications of a first born.

MARTYRDOM

קידוש ה׳

CHAPTER TWO

"O' Angels go forth, and with a bitter cry...
Lament for the congregation of scattered sheep
Together they assembled in holiness pure...
To sanctify the great and Holy name."[1]

Unfortunately, "kiddush hashem" (martyrdom) is a recurrent theme in Jewish history. Martyrdom was a choice many Jews willingly made. Throughout the ages, Jews suffered physical and spiritual torture; they endured cruel and inhuman pain; and they submitted to barbarous death sentences rather than forsake the tenets of Judaism. The martyrs asserted by their death that so long as they were not permitted to live as Jews, they would not live at all.

Some periods in which Jewish martyrdom bore eloquent testimony to religious conviction were the Hellenist period in the second century B.C.E. (before the common era), the fanatic repressions of the Hadrianic era in the second century of the common era, the forced conversions to Christianity during the Crusades in the eleventh through thirteenth centuries, and the merciless slaughter of Jews during the Chmielnicki pogroms in the middle of the seventeenth century.

The Rabbis who lived during the era of the Hadrianic persecutions were faced with the problem of martyrdom. Jews were martyring themselves when faced with the choice of violating all types of Jewish law. The Rabbis wanted to restrict the areas mandating the necessity to die "al kiddush hashem" based upon the biblical verse "and you shall love the Lord your G-d with all your heart, with all your soul and with all your might."[2] They interpreted this verse to be the prohibition against worshipping any other in His place, and therefore they mandated martyrdom in the case of forced worship of pagan gods or forced conversion to other religions. Under different circumstances, a Jew was expected to live by the laws of the Torah and not die because of them.

> "The Jewish court which met in the upper chambers of the house of Nitza in Lydda resolved by a majority vote that in every other law of the Torah, if a man is commanded 'transgress and suffer not death', he may transgress and not suffer death, excepting idolatry, adultery and murder."[3]

The stories in this chapter vividly portray the significance of Jewish suffering and the obstinate will of the Jewish people to survive.

1 Kinot Tisha B'Av: Hecherishu Vaadabayrah (paraphrased)
2 Deuteronomy 6:5
3 Talmud Bavli Sanhedrin 74 a

MASADA[1]

During the early centuries of the common era, the Roman empire ruled Eretz Yisrael. Throughout the long reign of these brutal foreign masters, martyrdom reached epic proportions in Jewish life. Whole communities committed suicide rather than submit to the harsh repressions imposed upon the Jewish people by the hated and tyrannical Romans.

Masada is the most famous of the many communities in which the inhabitants chose martyrdom rather than surrender to Roman superiority. The Jews who lived in Masada decided on mass suicide, because for them there was no alternative. They chose to die in freedom rather than to submit to tyranny. They chose the freedom of death.

Masada was a fortress which was built out of rock on a mountain in the Dead Sea region, southeast of Jerusalem. Shortly after the destruction of the second Holy Temple, a group of freedom fighters took refuge in Masada. These freedom fighters were determined to defy Rome's ambition to subdue the Jews of Eretz Yisrael who desperately hoped to retain their freedom through open acts of rebellion against the empire. The freedom fighters knew their situation was almost hopeless but they chose to die as free men rather than as Roman slaves or captives fed to the lions at the Roman circuses.

Masada had an abundance of food supplies to sustain the approximately nine hundred sixty inhabitants for a long period of time. There was ample wheat, oil and water to withstand a protracted siege. The freedom fighters used unusual battle strategy to defend themselves against the Roman soldiers who dared ascend the ramp to Masada. The freedom fighters poured boiling oil upon the heads of their attackers but this ingenious strategy caused only a temporary Roman retreat. When the freedom fighters recognized that they could not survive in the face of overwhelming Roman forces, they chose to commit suicide.

The leader of the freedom fighters at Masada was Eleazar ben Yair. He said:

> "'Let our wives die before they are abused and our children before they have tasted slavery: and after we have slain them, let us bestow that glorious benefit upon one another mutually and preserve ourselves in freedom as an excellent funeral monument for us. But first, let us destroy our money and the fortress by fire: for I am well assured that this would be a great grief to the Romans, that they shall not be able to seize upon our bodies, and not avail themselves of our wealth also; and let us spare nothing but our provisions, for they will be a testimonial when we are dead, that we were not subdued for want of necessaries, but that, according to our original resolution, we have preferred death to slavery.'

> "Afterwards, they chose ten men by lots, to slay all the rest; everyone of whom laid himself down by his wife and children on the ground and threw his arms about them and they offered their necks to the stroke of those who by lot executed that melancholy office. Then the ten remaining cast lots among themselves, that he whose lot it was should first kill the other nine, then should kill himself.
>
> "When Eleazar ben Yair perceived that they were all slain, he ran his sword entirely through himself and fell down near to his own relations."[2]

On the fifteenth day of the Hebrew month of Nissan, the first day of Passover, three years after the destruction of the second Holy Temple, the Romans investigated the awesome silence which had befallen Masada. The Romans expected resistance to their attack, but "a terrible solitude on every side, with a fire within the palace, as well as perfect silence greeted them. Nor could they do other than wonder at the courage of the defenders and the immovable contempt which so great a number of them had shown".[3]

Thus, on Passover, the festival of freedom, approximately nine hundred sixty people, men, women and children, chose the freedom of death by refusing to succumb to Roman tyranny.

1 Josephus Flavius: Antiquities of the Jews: War of the Jews, Books III, IV. Translated by Whiston. Bigelow & Brown Co. New York.

2 Ibid.

3 Ibid.

THESE I REMEMBER[1]

The Roman emperor Hadrian attempted to methodically strangle the life blood of the Jewish people by instituting more cruel prohibitions than previous emperors. He tightened the prohibitions against the study of Torah and decreed the cessation of Rabbinical ordination. The Jews answered Hadrian by declaring their stubborn determination to die "al kiddush Hashem", to sanctify G-d's Holy Name and His Torah.

The story of the "Asarah Harugay Malchut" (ten illustrious sages who died to sanctify G-d's Holy Name) is part of the liturgy recited twice each year, on Yom Kippur (the Day of Atonement)[2] and on Tisha B'Av (the ninth day of Av)[3].

It is a dramatic presentation of historical events that actually occurred, written in the form of "piyut" (prayer poem). The way the "piyut" reads, it seems as if the ten illustrious sages were martyred at one time. However Rabban Shimon ben Gamliel died approximately sixty years before during the period of the destruction of the Second Holy Temple, and Rabbi Ishmael ben Elisha Kohen Gadol died about twenty years before the story of the "piyut" actually took place.

The "piyut" begins with Joseph, his brothers and their descendants. Joseph, at Jacob's request, goes to investigate the welfare of his brothers while they are tending the sheep. The brothers, who are jealous of Joseph's dreams, want to kill him but they decide to sell him to a group of Ishmaelite traders instead. The Ishmaelites sell Joseph to a group of Midianites and they in turn sell Joseph to Potiphar, the captain of the Pharaoh's guard in Egypt. Gradually, Joseph rises to the position of Prime Minister of Egypt. He invites Jacob and his brothers and their families to join him in this new land where the Jewish people multiply from a family of seventy souls to a population of 600,000.[4]

Following Joseph's death the Jewish people are enslaved, but after many years of servitude, they are finally redeemed.[5]

Torah commentators explain that the story of Joseph and his brothers was G-d's plan for bringing the Jewish people into Egyptian slavery and molding them into a nation. The ultimate purpose of the slavery in Egypt and the redemption was the Revelation at Mt. Sinai.[6]

The Roman Emperor Hadrian used the story about Joseph's brothers which he found in the "Septuagint" (the Greek translation of the Torah) as a pretext to justify the condemnation of ten innocent Jewish martyrs. First, he gathered the ten leading Jewish sages and conferred upon them the power of a judicial court. Quoting the Biblical passage, "he who kidnaps a person and holds him for ransom will surely be put to death"[7], he presented a case against the Jewish community. He argued that no one had been punished for the kidnapping of Joseph centuries before. Without waiting for their verdict, Hadrian pronounced the sentence of death upon these ten sages, the descendants of Joseph's brothers who represented the Jewish people.

The ten Rabbis died for the sanctification of G-d's holy Name, unwavering in their belief that Torah was eternal.

Rabbi Akiva was condemned to die in Caesaria. It was early in the morning when Jews ordinarily recite Sh'ma that Rabbi Akiva was brought before the Roman tribunal. Unafraid, he continued to recite Sh'ma as his flesh was torn from his body with iron combs. His students asked him why he was smiling and continuing his prayers when the torture was unbearable. He answered:

> "I never knew the meaning of the sentence: 'You shall love the Lord Your G-d with all your heart, with all your soul and with all your might'[8] until I had to die to sanctify His Name."

As he was dying, Rabbi Akiva taught the Oneness of G-d and the survival of the Jewish people. His students buried his body in the cemetery outside the Galilean city of Tiberias.[9]

Rabbi Chananya ben Teradyon was studying the Torah when the Roman bailiffs came to escort him to his execution. His wife and daughter were weeping. He asked his daughter why she was weeping and she answered that she was weeping for the Torah that would be burned with him at the time of his execution.

He answered:

> "I see the parchment consumed with flame, but the letters of the Torah are returning to their source in Heaven."[10]

The Romans then insisted that lots be cast to decide who would be slaughtered next, Rabbi Yishmael ben Elisha Kohen Gadol or Rabban Shimon ben Gamliel. The lot fell on Rabban Shimon ben Gamliel who was immediately decapitated. As Rabbi Yishmael ben Elisha picked up the severed head he bitterly cried out that the tongue which was so skilled in transmitting the words of Torah was now licking the dust.

A royal matron was watching Rabbi Yishmael ben Elisha who was a very handsome man. She requested the executioners to ask him to raise his head so that she might gaze upon his comely features. Had he acquiesced, she was prepared to grant his life, but he refused saying that he would not forfeit his life in the world to come for her to have one hour of pleasure at his expense. She ordered that his skin be flayed from his face.[11]

Rabbi Yehudah ben Bava was killed by three hundred iron spearheads which the Romans pierced into his body. His crime against the Romans consisted of ordaining a new generation of Rabbis. He had gathered five of his most worthy students in the uninhabited valley between two mountains near the cities of Usha and Shefaram. He wanted to ordain them in an uninhabited area because the Romans had decreed death to the inhabitants and destruction to any city that allowed ordination to take place within its borders. The aging Rabbi, not wanting harm to befall innocent people, ordained his students in this uninhabited valley. As soon as he bestowed the title of Rabbi on these worthy students, he pleaded with them to flee swiftly for he knew the Romans were following him. At first the students refused to leave their revered teacher, but he, as well as they, knew that he was too old to escape the pursuing Roman soldiers. When the Roman soldiers caught Rabbi Yehuda ben Bava, they brutally murdered him.[12]

Rabbi Chutzpit Hameturgeman was one hundred and thirty years old when he was "sentenced" to die. His colleagues pleaded with the Roman emperor to have mercy on him. The emperor asked Rabbi Chutzpit how old he was.

The sage answered:

> "I am one hundred and thirty years old less one day and I beg you to give me this day in order that I might fulfill the commandment of reciting the Sh'ma this evening and tomorrow morning to acknowledge the Almighty as my King."

The emperor demanded to know how long the Jewish people would cling to a G-d who did not help them. Rabbi Chutzpit cried bitterly and rent his clothes in anguish as he listened to the pagan emperor blaspheme the Lord.

He cried out:

> "What will you do mighty emperor when the Lord punishes Rome?"

The emperor, hearing the audacity of the old sage commanded that he be slain, stoned and hanged immediately.

Rabbi Yeshayvav, the Scribe, was teaching his students to cleave to one another, love peace and justice and pray for redemption when he was taken to the place of execution. It was the second day of the week, the day on which he usually fasted. He refused to deviate from this custom. As he was being executed, a voice was heard saying:

> "Blessed be Rabbi Yeshayvav the Scribe who never for an instant faltered from the law of Moses."

Rabbi Chanina ben Chachinai was led to execution on Friday afternoon while he was finishing his preparations for the Sabbath. When he reached the site of the execution, he began to chant the Sabbath eve prayers, ushering in the day of rest. The executioner lingered until Rabbi Chanina reached the verse:
"And G-d blessed the seventh day and sanctified it...," then he acted. Rabbi Chanina never finished the remainder of the verse: "for on that day, G-d rested from all His work which He had created."[13]

Rabbi Yehuda ben Dama was executed the day before the feast of Tabernacles. He pleaded with the emperor to delay his execution until he had performed the commandments of sitting in a Succah and blessing the "etrog" and "lulav" (palm branch and citron). The emperor mocked Rabbi Yehuda's faith. He wanted to know the reward that Judaism promised the righteous.

Rabbi Yehudah answered:

> "Oh, how abundant is Thy Goodness which Thou has stored up for them that fear Thee."[14]

The emperor said:

> "You are a fool if you believe in life after death."

Rabbi Yehudah answered:

> "There are no fools greater than those who deny a living G-d."

When the emperor heard these words, he commanded that Rabbi Yehudah ben Dama be tied by his hair to a horse's tail and dragged through the streets.

Rabbi Eliezer ben Shamua was led to his execution on the Day of Atonement. He was a humble and gentle man who never quarreled with his colleagues even though they disagreed with his opinions. He never occupied himself with useless matters. His students pleaded with him to divulge what he saw in the last few moments prior to his execution.

He answered:

> "My children, I see the souls of every righteous man being purified in the waters of the Shiloah River, preparing to enter the gates of Heaven where Rabbi Akiva is already sitting upon a golden chair teaching Torah to them who enter."

The martyrdom of these sages reminds us that they believed in the eternity of the Jewish people and the Torah.

1 Ahron Jellinek. Bayt Hamidrash. "Midrash Ayleh Ezkerah". Bamberger & Wahrmann. Jerusalem. 1938.
2 Ayleh Ezkarah.
3 Arzay Halevanon.
4 Genesis 37-47
5 Exodus 1-15
6 Rashi, Exodus 20:2
7 Exodus 21:16
8 Deuteronomy 6:5
9 Talmud Bavli Berachot 61 b
10 Talmud Bavli Avodah Zarah 18 a
11 Midrash Ayleh Ezkerah
12 Talmud Bavli Sanhedrin 14 a
13 Genesis 2:3
14 Psalms 31:20

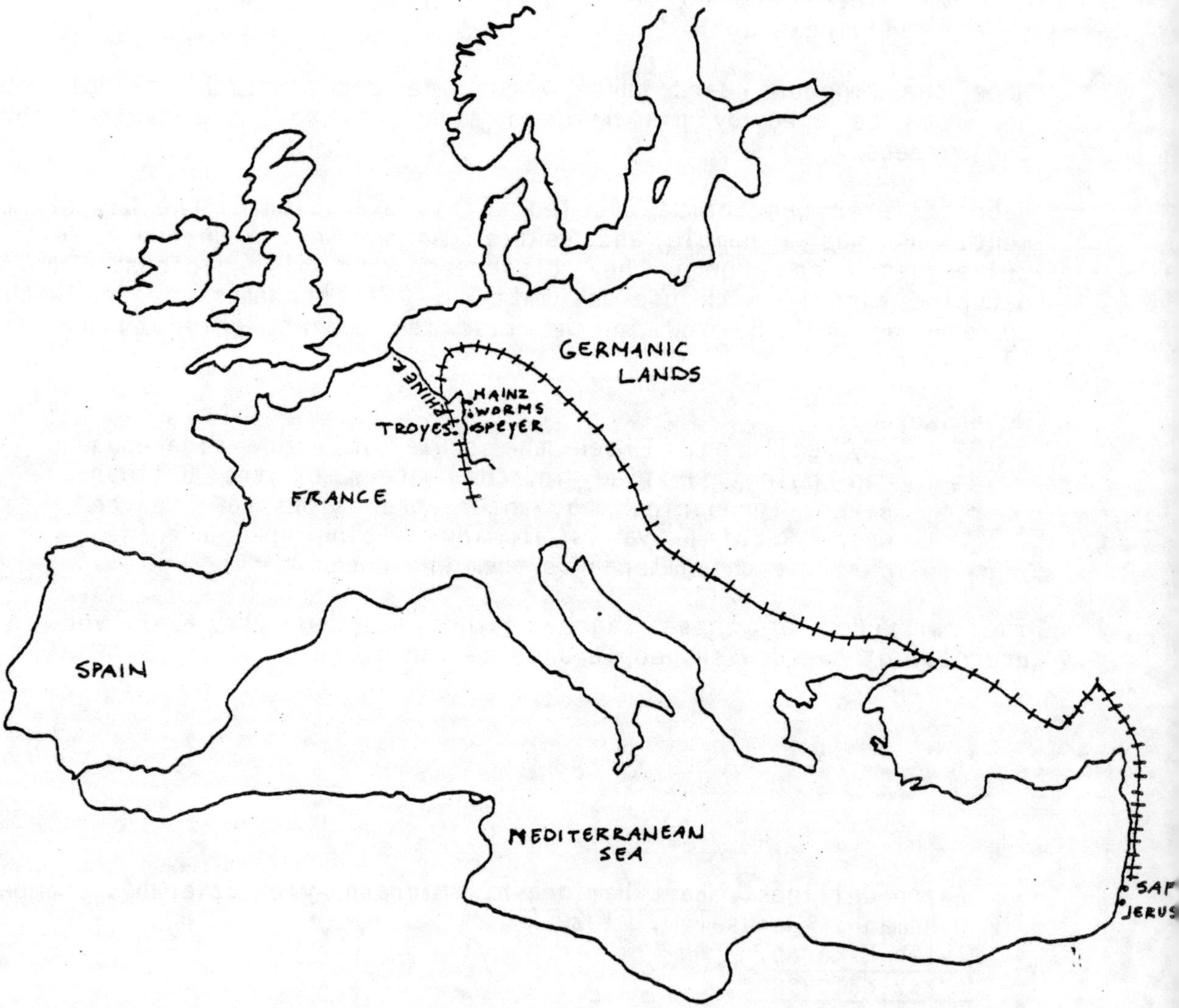

++++++++ ONE OF THE APPROXIMATE GERMAN CRUSADE ROUTES 1096

THE JEWISH POPE[1]

The two hundred year period, from the eleventh to the thirteenth centuries, known as "Maasay Hatzlav" (the Crusades) was a time when martyrdom was so prevalent that a special blessing "Mekadesh et shimcha barabim" (whose Name is Sanctified before the whole world) was formulated. Historians have recorded these years as "Gezayrot Tatnu" (the pogroms of the year 4856) (1096 C.E.). Thousands of Jewish victims of expulsion, pillage and murder chose to die rather than to submit to the forced conversions imposed upon "infidels" by the Crusaders who were on their way to free Eretz Yisrael from Moslem rule.

The degradation and horror which the Crusaders perpetrated against the Jewish people created a feeling of utter hopelessness. The remnants of the massacred Jewish communities could not revenge the atrocities inflicted upon them. The only spark of strength for survival the Jews possessed was their faith. They preferred to die as martyrs rather than renounce that faith.

The liturgy of that period is full of expressions describing the pain of horrible suffering. Incorporated into the prayerbook are lamentation prayers specifically written in memory of the martyrs. The prayer "Av Harachamim" (Our Merciful Father), recited every Sabbath morning, in many synagogues, is one of those lamentations. Many additional lamentations were incorporated into the Rosh Hashana (New Year) and Yom Kippur (Day of Atonement) prayerbook, some of which were written by the martyrs themselves prior to their death. Tradition suggests that the recitation of "Yizkor" (the memorial prayer for the deceased) recited during "Pesach" (Passover), "Shavuot" (Pentecost) and "Succot" (Tabernacles) began during the period of the Crusades when Jewish martyrs died for the sanctification of G-d's Holy Name.

Three examples portray the stubbornness of the Jewish people to survive the evil perpetrated against them.

One Sabbath morning, the Jews of the Speyer Jewish community who gathered in their synagogue to pray were surrounded by a legion of Crusaders. The Crusaders demanded that the Jews come out, renounce their faith and accept baptism. The Jews refused to heed the demands of their tormentors. The Crusaders bolted the doors and set fire to the synagogue, burning alive the praying Jews who died for the sanctification of G-d's Holy Name.

Worms was another decimated city. Its leader was Rabbi Eliezer. A parchment "Hagada" (the story of the Exodus from Egypt) beautifully bound and painstakingly illuminated is displayed in the National Museum in London in the Antique Manuscript And Rare Book Department. It was handwritten by Rabbi Eliezer and it is a remnant of the quality of Jewish life in Worms before the Crusaders destroyed that city.

The Jews believed that Jewish survival depended on their ability to preserve Torah. Even though the Crusaders prohibited the public reading of the Torah in the synagogue during the Sabbath morning prayers, the Jews ingeniously dealt with this restriction by continuing the ancient custom of reading the Haftorah instead.[2] The custom of reading the Haftorah in addition to the Torah originated during the period of Greek tyranny when the Jews were also prohibited from reading the Torah. Since the Crusaders did not realize the significance of the Haftorah, the Jews substituted the Haftorah for the Torah reading. They were temporarily able to outmaneuver their tormentors by reinforcing their belief in Jewish survival, albeit surrounded by adversaries.

This story is an historical legend about the Jewish past from the period of the Crusades.

Feudal lords were supposed to protect the Jews who lived in their territories in exchange for the taxes which the Jews paid them. During the Crusades, this protection was half-hearted and passive.

Rabbi Shimon Hazaken was the leader of the Jewish community of Mainz. He was Rashi's uncle, the brother of his mother. Because he was known as a great scholar, people travelled great distances to learn Torah and Talmud with him. He also wrote "piyutim" (prayer poems). Many of his prayer poems were incorporated into the "Machzor" (the Rosh Hashana and Yom Kippur prayerbook).

Rabbi Shimon had a five year old son whose name was Elchanan. The family employed a Christian governess to care for him. She always confessed to her priest that Elchanan was a very special child. Many priests encouraged their congregants to kidnap Jewish children in order to save their "souls" and this priest was no exception.

It happened several months before Rosh Hashana that Elchanan became very ill. Rabbi Shimon was very concerned about his son's recovery so he composed a prayer poem, pouring out his heart to G-d. The words had special significance because Rabbi Shimon composed the prayer poem in the form of an acrostic, and one special verse was composed with Elchanan's name. The name of the prayer poem is: "Melech amon maamorecha" (King, Your Word is steadfast) and the verse with Elchanan's name is "El chanan nachlato benoam lehashpar" (G-d has graced His people with the beauty of holiness). While he was recovering, Elchanan learned this verse by heart because he knew it was in his honor.

On Rosh Hashanah that year, Rabbi Shimon and his wife went to the synagogue, leaving Elchanan at home with the Christian governess because he was still recuperating from his illness. The house was empty except for the governess and the child. The governess knew that no one would return home for a long time. She had been waiting for this opportunity to execute the plot which the priest had instigated. She kidnapped the child from his bed and carried him to the priest. From the church, he was taken to a monastery in another city. The priest was rewarded by the bishop for "saving" a Jewish child for Christianity.

Imagine the anguish that Elchanan's parents suffered! They suspected the governess had kidnapped their son, but Jews could not do anything to defend their rights in those days. The Church was the government and the government was the Church.

Elchanan grew up in the monastery. He was very bright and therefore was privileged to receive an education. He became proficient in Latin, Greek and Hebrew, the languages by which all the knowledge of the Middle Ages was transmitted. The more he was exposed to Christianity, the more he forgot his Jewish childhood; yet Elchanan always remembered the one sentence of his father's prayer poem. Gradually, he was promoted through the hierarchy of the church, and when he was eighteen years old, Elchanan, son of Rabbi Shimon Hazaken, was appointed assistant to the Pope.

One day, Elchanan was riding in his carriage when he heard a Jewish girl screaming for help. She was being assaulted by two scoundrels. He jumped out of the carriage to assist the girl but the scoundrels persisted, advising him not to pay attention since this was "just a Jewish girl". Elchanan grabbed the girl, led her safely to his carriage and took her home. As he spoke comfortingly, he tried to assure her that he would protect her. The girl's father, one of the few wealthy Jews in the town, was very grateful to him for saving his daughter and he invited Elchanan to visit whenever the opportunity was convenient. Elchanan accepted the man's offer and visited the Jewish family many times, always incognito.

The Pope left instructions in his will that upon his death, his advisors should appoint Elchanan as his successor. When he died, Elchanan was elected to the Papacy. The son of Rabbi Shimon Hazaken took the name of Pope Victor III.

The parish priest of Mainz, who had followed the career of the child he kidnapped to the Papacy, travelled to Rome to see the new Pope. During an interview, he confessed that he was the instigator of the kidnapping and asked for financial reward.

> "Now that you are Pope," the parish priest claimed, "you certainly can pay me in measure for the great favor which I did for you, and for the great honor and wealth which the church has bestowed upon you."

The Jewish Pope was startled with this discovery. Without thinking, he gave the priest his reward. Then silently and alone, he went into the privacy of his chambers. His anguished heart convulsed with throbbing pain at the realization that he was a Jewish Pope.

The Pope decided that he had to find his father again, so he went to his Jewish friend for advice. His friend advised him to convene a conference of Europe's leading Jewish scholars, hoping that among the delegates in attendance would be the heartbroken Rabbi Shimon Hazaken. The Pope followed the advice of his friend and he sent out a decree ordering the heads of every Jewish community in Europe to attend a conference.

In the beginning, the Pope talked with the delegates; then he asked the great Rabbis to exchange religious ideas with him. Rabbi Shimon Hazaken and the Pope developed an amicable relationship during the conference and when it was over, the Pope pleaded with Rabbi Shimon Hazaken to remain in Rome to study with him for three months. After that time, Rabbi Shimon Hazaken decided to return to Mainz. As they were bidding farewell to each other, the Pope asked the great Rabbi how his family fared during his lengthy absence, and for unexplainable reasons Rabbi Shimon Hazaken remembered the prayer poem he had written for his son. He began to recite the phrase:

> "El chanan benoam ..." (G-d has graced His people with the beauty of holiness).

The Pope recognized the verse and completed it with the word "lehashpar". Both father and son immediately recognized each other. Rabbi Shimon Hazaken pleaded with his son to follow him to Mainz to embrace Judaism again. The young Pope hesitated to leave the kingship of the Church because he thought he could do many great things to ease the suffering of the Jewish people.

The Pope was known for his kindness, patience and wisdom, and the Christian people hoped he would rule for a long time.

One day, the Pope disappeared from his palace. The Church Council said the Pope went straight to heaven, but he really went incognito to stay with his Jewish friend. There, he married the girl he had saved and together they returned to Mainz. They pretended they were distant relatives so that none would know their true identity.

A few years later, on the second day of Rosh Hashana, the Crusaders were marching through Mainz, burning, plundering and raping in their path. The Jewish community was given the choice of conversion or death. The Jews refused conversion; so the entire community was marched to the market place to be burned to death.

Elchanan, Rabbi Shimon Hazaken's son, the ex-Pope, tried to interfere in order to save his people. He cried out:

> "Who knows the most about Jews? Who knows most about Christian subjects?"

The leader of the Crusaders answered:

> "The Pope!"
>
> "I am your Pope!" screamed Elchanan. "Stop killing the Jews! There is one G-d!"

Then Elchanan jumped into the fire. The prayer poem which Rabbi Shimon Hazaken wrote for Elchanan's recovery is recited on the anniversary of his death, the second day of Rosh Hashana.

1 Y.L. Baruch. Sayfer Hamoadim: Rosh Hashana V'Yom Kippur. Vol. I. The Oneg Shabbat (Ohel Shem) Society. Dvir Publishing Co. Tel Aviv. 1952.

Aḥron Jellinek. Bayt Hamidrash. Bamberger & Wahrman Publishing Co. Jerusalem. 1938.

2 The Haftorah is a selection from the prophetic writings which relates to the Torah reading assigned for a particular Sabbath by similarity of content or historical connection.

I WILL DIE A JEW[1]

This story about Rabbi Amnon of Mainz is so popular that it is often included in the High Holiday Prayerbook as an introduction to the prayer "U'netaneh tokef, kedushat hayom".

Leaders of the Christian Church preached that Christianity was a religion of love yet they encouraged the Crusaders to destroy Jewish communities that happened to be in their path. One of these communities was Mainz, a city that was famous for its "yeshiva".

Rabbi Amnon was a man of distinguished ancestry who lived in the eleventh century in Mainz. He was friendly with the Bishop of Mainz. The two friends enjoyed each other's company, and they spent many hours discussing intellectual matters. It was the Bishop's custom to urge Rabbi Amnon to convert to Christianity at the end of each visit. Rabbi Amnon always adamantly refused, but one time, under pressure from both Crusaders and the nagging Bishop, Rabbi Amnon consented to consider the matter. He told the Bishop he would let him know his decision in three days.

Rabbi Amnon realized that he had made a serious mistake as soon as he said he would consider the matter. He had never hesitated in his answer to the Bishop before. He shuddered at the thought that he had left the Bishop with the mistaken impression that he would convert to Christianity. He decided he would not return to the Bishop.

When Rabbi Amnon did not appear before the Bishop on the third day, soldiers were sent to remind him of their appointment and to escort him to the Bishop's home. Rabbi Amnon refused to go with the soldiers; so they dragged him through the streets of Mainz. Then they brought him before the Bishop.

The Bishop threatened Rabbi Amnon with severe punishment, but Rabbi Amnon pleaded to decree his own fate. He asked the Bishop to cut out the tongue that dared to speak about conversion. The Bishop was not satisfied with the suggestion of self-imposed punishment. Instead he ordered that Rabbi Amnon's arms and legs be dismembered and that salt be poured on the wounds to inflict greater suffering.

Rabbi Amnon begged his faithful followers to carry his bleeding and dismembered body into the synagogue and to place it on the pulpit. He was writhing in pain. As he lay dying in martyrdom, he affirmed his faith in G-d. He desperately wanted his followers to know the misfortunes which might befall man and his precarious position in the world. Yet he believed that faith and the acknowledgement of G-d's sovereignty shielded man from succumbing to despair. Rabbi Amnon wanted the world to know that the Jewish people acclaimed the Oneness of a living G-d.

Rabbi Amnon interrupted the "chazan" (prayer leader) who was in the midst of chanting the Musaf prayers, the additional prayer service for the Rosh Hashana festival. He began to recite the words of the "U'netaneh tokef kedushat hayom".[2]:

"May our sanctification of Thy Name ascend,
For You are our G-d and our King.
Let us observe the Holiness of this day.
This is the day Your Kingdom is exalted,
It is established in mercy and truth.
You are the Judge of mankind...
And You hold each man accountable for his deeds...
On Rosh Hashana day man's destiny is inscribed
And on Yom Kippur his fate is sealed...
Who will live and who will die...
Who by fire...who by water...
Who will be humbled...who will be disturbed...
But repentance, prayer and charity avert an evil decree...
And You are Our G-d, Our Living King."

As Rabbi Amnon uttered these holy words, his soul departed. Three days after Rabbi Amnon's death, Rabbi Kalonimus ben Meshullam dreamed that Rabbi Amnon was teaching him the words of "U'netaneh tokef". In the dream, Rabbi Amnon commanded him to travel to the Jewish communities in the Diaspora in order to teach this prayer.

Rabbi Kalonimus followed the command in the dream and he began to travel. Wherever he went, he taught Rabbi Amnon's prayer. To this day, it is recited on both days of Rosh Hashana, the anniversary of his death and on Yom Kippur.

1 Eliyahu Ki Tov. Sayfer Hatodaah. Yeshurun Publishers. Jerusalem. 1966

2 Paraphrased and translated from Machzor: High Holiday Prayer Book. "U'netaneh Tokef Kedushat Hayom".

SHIMON THE SILENT[1]

Around the thirteenth and fourteenth centuries, the Jewish ghetto[2] became the address of Jews living in every developed country of Europe. The ghettos were separate sections of villages that were surrounded by gates and walls. They were usually in the slum section of the village where the houses were dilapidated and situated on narrow, winding streets. There was no room for expansion in these confined areas, and as the Jewish population increased, the dwellings became overcrowded. Living conditions grew progressively worse.

The Jews were treated as second class citizens by rulers, religious leaders and the masses in Europe. They were subjected to the indignity of sleeve badges and special hats to reveal their identity to the world. Volumes of the Talmud were publicly burned and writing new holy books was prohibited. They were accused of exploiting the townspeople through economic competition. They were generally excluded from the trade and commerce guilds which controlled both production and distribution. They could eke out a living doing only the most undignified tasks. Jews were not permitted to hold public office, to employ non-Jewish servants, to join professions, to convert Christians to Judaism, to intermarry or to own land. The Jews were continuously subject to edicts of expulsion, and they could find new places to live only where a ruler might benefit from an exorbitant entrance price or oppressive taxes.

At first the Jews complained to the authorities about living conditions in the ghetto and the economic restrictions placed upon them, but to no avail. The Jews had to learn to live under ghetto conditions which inadvertently benefited them by creating strong, cohesive, independent ghetto communities whose walls protected them from the infiltrating mobs on the outside. Prague, Czechoslovakia was one of these strong, independent ghetto communities.

The Jewish settlers were permitted to build a synagogue, and Jewish life in Prague was centered around the "alt-neu shul" (the old-new synagogue), the oldest remaining Jewish house of worship in Europe today. The "alt-neu shul" was so significant to the Prague Jewish community that the people believed it bridged heaven and earth in Prague much as the "Kotel" (the Western Wall) bridged heaven and earth in Jerusalem.

This story is about the martyrdom of one individual. Had Shimon the Silent, the hero of this story, not chosen personal martyrdom, the entire Prague Jewish community would have been massacred.

A tailor named Shimon, a man of very few words, lived in the Prague ghetto. The people called him Shimon "the quiet one" or Shimon "the silent". Shimon was the object of mockery and scorn.

The only people who were disturbed by the mocking were Shimon's wife and mother. His wife was grievously distressed by it but his mother, the caretaker of the women's gallery in the "alt-neu shul" always rose to his defense.

Once near the holiday of Shavuot (Pentecost), the inhabitants of the ghetto received a special message that the King was coming on an inspection visit. The Jews gave their allegiance to the King since peaceful life in the ghetto depended on his protection. The Rabbi urged the community to greet the King with great honor. The houses and streets in the ghetto were cleaned and flowers were arranged to adorn the marketplace.

The King arrived in the ghetto at the appointed time and he inspected it carefully, pleased with what he saw. Suddenly, from the top of the roof, a sharp stone was thrown at the King's head aimed at murdering him. The stone missed the King's head by one inch. The Jews were certain that it was not one of them who did this terrible deed but it was impossible to convince the King of their innocence. The King left the ghetto furiously and decreed that if the murderer were not found within eight days, he would allow a mob to carry out a pogrom against the entire ghetto.

The eighth day was Shavuot (Pentecost), the holiday which celebrates the receiving of the Torah on Sinai. The anticipated joy associated with this festival was quickly becoming a time of sadness and fear for the worst.

No stone was unturned to find the murderer but there was not even a clue. The holiday was drawing closer. The Jews knew that the King would not rescind his decree, and the mob waited with wild excitement outside the ghetto. The Jewish community gathered sorrowfully in the "alt-neu shul" to pray. There was nothing they could do to prevent the pogrom.

On the morning of the eighth day, Shimon said goodbye to his wife and children and disappeared. No one noticed that Shimon "the Silent" was not present in the "alt-neu shul" for the holiday prayers. Only Shimon's mother realized his absence and she searched all over for her son but she could not find him.

Suddenly there was a knock at the ghetto gates. With great trepidation, the Jewish people emerged from the "alt-neu shul", slowly walked toward the gates and opened them. A messenger read a proclamation from the King stating that the murderer had been found and the evil decree was withdrawn. The messenger continued that the murderer would be thrown off the same roof from which the stone had been thrown. Soldiers would be stationed on the ground beneath the roof with outstretched bayonets to stab the falling body of the confessed murderer.

The Jewish people were forced to watch the execution. When the murderer was pushed to the edge of the roof, the horrified people gasped as they looked upon the trembling face of Shimon "the Silent".

His mother, realizing that her son had saved the Jewish community, ran inside the "alt-neu shul" and opened the holy ark. She offered a thanksgiving prayer for having been chosen to be the mother of a holy son who was ready to sacrifice himself just as our patriarch Isaac was willing to do. Just as the soul of our matriarch Sarah departed immediately after she heard the news of the binding of Isaac, so Shimon's mother's soul departed.

A few months later, the Mayor of Prague, Vladislov von Rosenberg, lay dying. He requested his servants to call the Rabbi of the Prague ghetto so that he might speak to him before he died. The Rabbi arrived at the spacious, glittering mansion and the mayor confessed that he had hired someone to throw the stone that had threatened the King's life. The mayor asked the Rabbi to beg forgiveness from the entire Jewish community.

The Rabbi answered:

> "I don't have the power to forgive but you can make amends so that you won't be deprived of the world to come. You can leave all your fortune to the children of Shimon "the Silent" so that they won't suffer so much in this world."

The mayor accepted the Rabbi's suggestion and left instructions in his will that his fortune be left to the children of Shimon "the Silent". The Jews of Prague realized that the man they had mocked had given his life to save them.

1 Sippurim Prager Sammlung: Judisher Legenden. R. Lowit Verlag. Wien und Leipzig. 1921. This book is an anthology of stories about the Prague ghetto. It is part of the literature collection in the Jewish State Library Museum in Prague. I discovered it when I explored the remnants of Jewish communities of eastern Europe in July, 1982. The author is unknown.

2 The origin of the word "ghetto" probably is the Italian form of borough, "borghetto", and it means the special quarter of a town or village where Jews lived. The ghetto in Venice, Italy, was located near the iron foundry, the "geto" in the early fifteenth century.

A CONVERSO BISHOP CELEBRATES PASSOVER[1]

The Spanish Inquisition was a special, permanent tribunal appointed by the medieval Catholic church to investigate and combat heresy.

The infamous Torquemada, chief religious advisor to the Spanish King Ferdinand and Queen Isabella, was the overseer of the Spanish Inquisition, and he was determined to protect Catholicism from being undermined by heretics.

Restrictions against Jews in Catholic Spain were enacted gradually. First, artisans and craftsmen found markets closed to their goods, then Jews were forced to wear coarse looking clothing decorated with a distinctive badge. Unwarranted attacks on life, limb and property became commonplace. Many Jews, realizing that the centuries of peaceful settlement had passed, fled Spain and resettled in countries that were more hospitable, notably the Netherlands and Poland. They fled by pretending to take their families on a vacation or business trip.

However, some Jews who did not want to leave Spain, chose to convert overtly but they remained hidden Jews in the privacy of their homes. These Jews were called "Anusim" (forced to accept another faith) or Conversos. The Church called the Anusim "Marranos" (pigs) or those who marred true faith, and considered them to be heretics. They became victims of many different barbaric methods of torture.

Life imprisonment was reserved only for the repentant heretic, but capital punishment, auto-da-fe, (burning at the stake) was meted out to those who refused to confess. In the course of time, the cruelty of the Inquisition surpassed all the preceding medieval forms of anti-semitism as one Jew after another was led to be burned at the stake.

This story describes an encounter between a converso bishop and two Jews, Chaim and Moshe, who escaped from Spain. They settled their families in Venice, Italy, and decided to risk their lives and return to Spain for business about two months before Passover. They were both disciples of Rabbi Shimon ben Yitzchak Simcha Luzzato.

On the boat trip from Venice, Chaim and Moshe discussed necessary measures to insure their safety while they were in Spain. They decided to part company at the dock and conduct their business separately, not communicating with each other until they met at a prearranged time and place for their return passage home. When the friends arrived at the harbor on the appointed date, they were informed that the boat on which they had booked passage was delayed a few days, making it impossible for them to return to Italy in time for the Passover holiday. They panicked at the thought of spending Passover in Spain, the country where the Inquisition was rampant.

The holiday was rapidly approaching, and Chayim and Moshe went trembling to the marketplace to buy some provisions to sustain them for the duration of their forced stay in Spain. They thought that if they bought bitter vegetables and wine, spies, placed by the

Inquisition in the marketplace, would inform on them to the authorities. They feared they might be arrested on suspicion and end up at the auto-da-fe. To play it safe, they decided to separate for the second time.

Since the situation was so dangerous, Moshe decided to buy some meat, potatoes and apples, find a room in an inn and stay behind closed doors for the entire week of the Passover holiday. He improvised a solitary "Sayder" and spent a lonesome festival.

Chayim, however, decided to pray to G-d that a sign would be revealed to him which would bring a solution to his problem. He waited at the edge of the marketplace. As the sun slowly set, he noticed a caravan of horsemen riding through the marketplace, led by a Bishop.

> "Buy up the market, load up the wagons," the Bishop ordered his servants, "tonight we will feast!".

Chayim believed that this was his sign, so he hid patiently and waited until the wagons were loaded with all of the goods from the market. As the caravan left the marketplace for the Bishop's mansion, Chayim ran after it. When the caravan reached the Bishop's mansion, the servants began to unload the food from the wagon. Chayim waited for a while outside the gate; then he banged on the door, shouting:

> "I must confess! I must confess!"

The servants decided the shouting man must be a new Christian who found the urgent need to confess at this strange hour in the early evening. They decided to let him in against orders. They opened the gate and ushered Chayim into the Bishop's confession chambers, requesting him to wait. The Bishop arrived and entered his usual hidden spot in the confession booth.

> "I'm listening," the Bishop said, "please begin".

Chayim, his heart pounding, hoping that he had not misinterpreted the sign which led to his decision to follow the Bishop, placed his hands across his face and whispered:

> "Sh'ma Yisrael Adonay Elohaynu Adonay Echad" (Hear O Israel, the Lord Our G-d, the Lord is One)[2] and he waited.

If the Bishop were not a Converso, a hidden Jew, he knew that he would be brought to trial, tortured and sentenced to death, but the Bishop flung open the door of the confession chamber and embraced Chayim.

> "We have been waiting for you," he quietly said, "follow me".

Chayim followed the Bishop as they descended into a long corridor. At the end of the corridor, there was a hidden door which opened to a secret room. The Bishop's family waited inside seated around a splendid table, set with sparkling dishes and special food prepared in honor of Passover. What a "Sayder" feast! What a Passover celebration! Each holiday symbol was a gem. Each song which the family sang was a treasured remembrance. Each moment that Chayim shared with the Converso Bishop and his family was precious. Chayim spent the most joyous Passover of his entire life as the guest of the Bishop.

When the festival was over, Chayim returned to the harbor where he had arranged to meet Moshe. They were both extremely anxious to leave Spain, but they hesitated to exchange greetings until they were safely aboard the boat. When the boat left the dock and it had departed from Spain, they exchanged their Passover experiences. Puzzled by the vast difference of the holiday celebration they experienced, they decided to ask their Rabbi for an explanation upon their arrival home.

Rabbi Shimon Luzzato questioned them both about their commitment to observing the laws of "Kashrut" (the dietary laws). Moshe explained that the laws of "Kashrut" were valid. However, he felt it was not imperative if he had no choice but to partake of non-kosher foods. Chayim explained that the law was his life and his faith, and his belief in G-d's word was more important than his appetite to eat prohibited foods.

The Rabbi explained that their experience was a test of their faith. Chayim who never ate non-kosher food under any circumstances merited the celebration of a holy Passover, but Moshe who was willing to compromise merited only a "compromised" holiday.

This story might have had a different ending. Had Chayim misinterpreted the sign which led him to the Bishop, he would have been arrested, tried and brought to Inquisition justice at an auto-da-fe.

1 This story was told by Rabbi Velvel of Brisk, the Brisker Rav, to his students. He died in 1938. One of his students retold it to Rabbi Shlomo Carlebach who retold it to me.

2 Deuteronomy 6:4

HABATKO, THE COSSACK CAPTAIN AND THE DAUGHTER OF THE RABBI[1]

Poland had been a haven for Jews who migrated from western Europe for approximately three centuries when Bogdan Chmielnicki assumed the leadership of the Cossacks, the undisciplined peasant hordes that inhabited the eastern plains of the Ukraine. They had no allegiance to the Polish King or the land, and they earned their living by hunting and plunder. They ravaged the countryside, burned towns and villages and massacred both Poles and Jews in their attempt to revolt against Polish political, economic and religious exploitation. The massacres which occurred in 5408 and 5409 (1648-1649 C.E.) are called "Gezayrot Tach v'Tat".

The Jews were the primary victims of the barbaric Cossacks. They were the lessees of agricultural land, forests, mills and inns. In this capacity, they collected taxes for the King and the nobles. Since the Jews were the tax collectors, they were the object of the Cossack peasants' wrath. For three long years, the Jews of Poland faced this tremendous calamity. Another chapter of Jewish martyrdom was added to the already tear-stained pages of Jewish history.

Entire cities were decimated. Nemirov, a large city with a wealthy Jewish population was penetrated by the Cossack peasant hordes through trickery. They posed as Polish soldiers; they dressed in Polish uniforms; and they carried the Polish flag. The Jewish gate-keepers assumed that the Polish King had sent troops to protect Nemirov from Cossack assault; so they opened the gates. They did not realize that the imposters had tricked them until it was too late to save the Jewish community from massacre.

This story is about the martyrdom of one Jewish girl who happened to be the object of a Cossack captain's affection. His name was Habatko. Prior to becoming a Cossack, he was a harshly treated peasant. However, this mistreatment did not affect his personality. He was gentle and different from the rest of the Cossack marauders.

He refused to participate in the atrocities of the pogroms against the Poles and the Jews. He was in love with a Jewish girl. She had the prettiest black velvet lashes; her eyes twinkled like the brightest stars, and her smile made him drunk with joy. He spent most of his time dreaming about her.

He met this young girl as he watched her cross a field. When he approached her, he told her that he loved her and asked her to become his wife in spite of his being a Cossack. He told her that she would have to convert to Christianity before they could marry. He promised her a silk purse and a golden crown as wedding gifts.

The young girl replied that she would rather die than marry under these conditions.

> "Yesterday," she cried, "both my parents were murdered. How can I, a Jewish orphan, ever find happiness in marrying a Christian Cossack?"

Habatko ignored her cries and refused to accept her excuses. He said quietly:

> "I expect that you will consent to marry me within two weeks. In the meantime, I will wait for you."

After two weeks, Habatko returned to the young girl. When she saw him, she offered him a wonderful gift that she had inherited from her mother in lieu of marrying him. She told him the gift was a magic pistol that could scatter bullets in a burst of fire:

> "It will protect you from any Polish assailants, she said."

Habatko laughed.

> "I don't believe that there exists such a gun that can accomplish what you claim."

The young girl replied:

> "If you don't believe me, put it to a test. Please test my gift right now and you will be convinced that I am telling the truth."

Habatko hesitated but he wanted to make her happy; so he loaded the gun and aimed it at the "mezuzah" on the doorpost, deliberately bent on destroying a holy Jewish object. At that moment, the young girl jumped to the doorpost and when the gun was discharged it killed her instantly. She fell beneath the "mezuzah". He ran over to the doorstep and kissed her realizing the trickery of his beloved who chose to die "al kiddush hashem" rather than forsake the tenets of Judaism.

1 Saul Tschernikovsky: Mivchar Shaul Tschernikovsky. Dvir Publishing Co. Israel. 1964.

THE RIGHTEOUS PROSELYTE[1]

Lithuania was the site of the great "Yeshivot" of the seventeenth century. They expanded with the northward migration of Polish Jewish refugees who had escaped the Chmielnicki massacres. Lithuania became a vibrant center of Jewish life, the pride of eastern European Jewry. The city of Vilna, a major Lithuanian city with a large Jewish population, was a citadel of Talmudic learning. In the Jewish section of Vilna, the sing-song melody of Talmudic study reverberated in the air in harmony with the sounds of the chirping birds. It was called the "Jerusalem of Lithuania".

The setting of this story, a famous part of our folklore, is in Vilna, the citadel of great Talmudic scholarship, but the theme is about "Kiddush Hashem".

Graf Potocki was the gifted scion of a celebrated aristocratic Lithuanian family who converted to Judaism. The Potocki family earned their money in a very lucrative liquor business. However, the family's manners seemed to resemble those of peasants rather than aristocrats.

One day, Countess Potocki went for a walk with her baby. The baby started to cry hysterically. She shook the baby carriage, trying to quiet her screaming child. She could do nothing to comfort him.

A Rabbi was walking on the opposite side of the street with some of his students. When he heard the bitter cries of the infant, he turned to his students, told them to wait for him and crossed to the opposite side of the street. He gently approached the mother and stopped to look inside the carriage at the crying child. Instantly, the baby stopped crying. The Rabbi turned to the mother and said:

> "Be patient with your child. I foresee for him a great future in the world."

As the boy grew up, he realized that there was more to life than raucous behavior, cursing and drinking, so he began to search for something other than material gratification.

When he reached his teen years, his parents bought him a horse. The boy spent a lot of time roaming the countryside on horseback. Sometimes he spent many days at a time roaming through the forests.

One Friday night as he was roaming through the outskirts of the Jewish part of town, he heard a beautiful melody coming from one of the smaller houses; so he tied his horse to a nearby tree and wandered over to listen to the melody. As he stealthily peered inside the window, he noticed that this house was much different than his luxurious home. There was a shining light radiating from the Sabbath candles that spread a soft glow throughout the house. The food on the table consisted of a challah and a small herring, but in spite of the meager meal, the family that sat around the table was joyously participating in Sabbath songs. He stood peering through the window, filled with a sense of peacefulness. He wanted so desperately to share the serenity of a family welcoming the Sabbath, but he only watched.

He returned to this house, week after week, hoping that he would eventually have enough courage to knock on the door and ask to be invited into the house. Finally, one Friday night, he knocked on the door. The people were very frightened to see a strange young man at their doorstep. The young man assured the family that he would not hurt them. He told the family that he had been watching them rejoice in the Sabbath meal; that he was trying to get up enough courage to knock on the door and invite himself into their home. The family assured him that he was welcome. They invited him to return the following Friday night. He did so and they continued to invite the young man, not only for Friday night, but for the entire Sabbath day. Gradually, the young man realized that his search for spiritual gratification would end if he could convert to Judaism, so he began discussing with the Jewish family the possibility of conversion. They warned him that Christianity frowned upon conversion to Judaism, and his actions might endanger the Jewish community. Realizing the sincerity of the young man to be converted to Judaism, they devised a safe plan for his conversion.

The Jewish couple suggested that the young man make plans to disappear from his family. They reasoned that it would take many days for them to realize that he was missing because he oftentimes roamed for days at a time. They thought that the young man's parents would assume that he met with a fatal accident on one of his outings if he did not return. In the meantime, he would leave immediately for Holland where he could convert without endangering himself or the Jewish people and pursue his Jewish studies without hindrance.

The young man followed the plan and went to live in Holland. He remained there for five years. During the years he was in Holland, he studied Torah, Talmud and commentaries. He learned quickly and achieved the reputation of being a "learned Jew".

After five years, the "learned Jew" decided to return to the Jewish family in Vilna. His appearance had completely changed from the rich young man who had decided to convert to Judaism. He had grown a beard and "payot" (sidelocks).

After spending some time with the family, the "learned Jew" took a liking to one of the daughters and married her. They had a number of children. The "learned Jew" enjoyed praying each Sabbath morning in the neighborhood synagogue. Knowing that the synagogue was a place of prayer, he resented the talking of the worshippers during the prayers. One Sabbath, he gently chided the worshippers for their bad habit of gossiping during the prayers and for their lack of decorum.

As he was gently chiding the worshippers about the respect due a synagogue, a boor shouted from among the congregants:

"Don't listen to him! Do you know who he really is? He is a convert. How dare he tell us what is right or wrong!"

The people were aghast at the discovery that the "learned Jew" was a convert. Word spread quickly through the city that the Jews were hiding a Christian in their synagogue. An army officer arrived on the scene and arrested the "learned Jew". He was taken to the army barracks where his beard and payot were shaved. His mother came to the barracks and identified him as her son. She pleaded with him to renounce Judaism and confess that the way she raised him was the only true way; but he refused.

The "learned Jew" was tried for heresy to the church, and he was condemned to be burned at the stake. When the morning of the execution arrived, an angry mob gathered in the marketplace to see the spectacle of yet another Jew giving up his life for the sanctification of G-d's Holy Name.

The execution of the "learned Jew" took place on the second day of the festival of Shavuot (Pentecost) in the Lithuanian city of Vilna. The last words of the "learned Jew", the righteous proselyte, Graf Potocki, were:

"Blessed art Thou, O Lord our G-d, who sanctifies His Name before the multitudes."

A local Jew, Eliezer Ziskes, pretending to be a Christian, bribed the local officials for permission to gather the ashes and a finger from the corpse of the "learned Jew". He buried them in the local Jewish cemetery. For years, the Jews of Vilna recited Kaddish at the grave of Graf Potocki every Tisha B'Av (ninth day of Av), Rosh Hashana (New Year), and Yom Kippur (Day of Atonement) to honor his memory. The reason they chose these particular holy days was because many themes of the prayers of these days remind us of those Jews who died "al kiddush Hashem".

1 I.M. Dick. Gayray HaTzedek. Vilna. 1862.

MOSHELE GOOD SHABBOS[1]

The years of World War II were a nightmare for world Jewry. Hitler rose to power out of the ashes of German defeat. He rekindled the flame of German national aspirations by making ashes out of untold numbers of innocent human beings.

Hitler was determined to annihilate the Jews and he proceeded in his determination with the words which a previous Jewish enemy used:

> "I will pursue; I will overtake; I will divide the spoil; my lust shall be satisfied upon them; I will draw my sword; my hand shall destroy them."[2]

The memories of the six million who died "al kiddush Hashem" bear witness to the painfully silent world which stood by while our people were destroyed.

So many words have been written about the Holocaust, but it is impossible to comprehend the meaning of the loss of the six million. Only the retelling of some of the stories that emerged from the Holocaust can begin to help each of us understand the real meaning of this great Jewish tragedy.

This story describes the last days of Jewish life in Vienna, in 1938, before the "Anschluss" (forced annexation of Austria to Germany). It is about "Moshele Good Shabbos", a Jew who never stopped singing his special Sabbath melody.

The ominous sound of the goose-step reverberated in the streets of Vienna, one of the most beautiful cities in the world.

The day was the Sabbath. The Jews wanted to pray with a "minyan" (a quorum of ten males) yet they feared to leave their homes. The "minyan" decided to gather before daybreak in order that they might finish praying before daylight so they could avoid the stormtroopers and return to their homes safely. They gathered to pray in a small house. They closed the doors and drew the shades so that they wouldn't be discovered.

Suddenly, a stark knock interrupted the whispered prayers in the small house where the "minyan" gathered to honor the Sabbath. A hushed silence spread through the house. A young boy ran to open the door. He was relieved when he saw Moshele Good Shabbos instead of a Nazi stormtrooper.

In a barely audible voice, Moshele sang:

> "My name is Moshele, what is your name?"

The young boy who had opened the door answered with the same sing-song melody:

> "My name is Shlomo, Good Shabbos, Good Shabbos."

Together, Moshele and Shlomo entered the room where the "minyan" had continued praying. They glanced at him momentarily, not wanting to be distracted from prayer. They were relieved to see one of the "most wanted criminals on the Gestapo traitor list", rather than an SS trooper. Moshele was listed because he forged passports to help Jews escape from Vienna.

Moshele continued to sing his special Good Shabbos melody as he wrapped himself in a "talit" (prayer shawl) and prepared to join the "minyan". He walked to the side, raised the drawn shades and opened the windows. In his same sing-song melody, he stuck his head out of the opened window and proclaimed:

> "Today is the Sabbath and we have to tell the whole world that all men are brothers. Good Shabbos, Good Shabbos."

He continued his same sing-song melody as he walked back to the center of the room where a man was leading the prayers in a whisper. Moshele said to him:

> "This is not the way to lead the Sabbath prayers. Today is a day of joy. We must shout Good Shabbos to the whole world."

The man asked Moshele to take over since he was too frightened to continue leading the prayers.

Moshele took over with the same sing-song melody as he sang his special Good Shabbos song. He began:

> "Nishmat kol chay" (every living soul blesses Your Name).

When he finished leading the prayers, he continued with his Good Shabbos, Good Shabbos melody.

The next day, Sunday, was the last day that Jews could leave Vienna if they had the proper documents. Moshele and his family were among those who were seated in a compartment on the last freedom train from Vienna. Nazi SS troopers stalked the train platform, searching for illegal escapees.

Moshele was overcome with emotion. He could not leave Vienna without saying goodbye. His wife and children pleaded with him to remain silent, but Moshele opened the window of the compartment, stuck his head out, and began to sing with his same sing-song melody:

> "Goodbye, Vienna, city of my birth, goodbye Vienna, city that I love."

Immediately two storm troopers recognized him, rushed into the train compartment and pulled him from the train. They began beating him unmercifully, not because he did not have the proper forged documents but because they knew he had helped others escape. All the while he was beaten, Moshele kept singing "Good Shabbos", until his last breath.

Moshele's survivors live in Antwerp and in Benay Brak, Israel, and they are still singing "Good Shabbos, good Shabbos".

1 Two people have corroborated the veracity of the following story: The little boy in the story is Rabbi Shlomo Carlebach, the Chassidic singer and story-teller, and the Kapishnitzer Rebbe, who lives in Benay Brak, Israel, is Moshele Good Shabbos' surviving brother. A son of Moshele Good Shabbos is living in Antwerp. Since he was so young during the Holocaust, he did not remember his father's melody. Recently, Rabbi Shlomo Carlebach was privileged to find the son of Moshele Good Shabbos and taught him his father's melody.

2 Exodus 15:19

THE HINKEDIKER TAILOR'S PLAN[1]

There were Jews who resisted and Jews who succumbed. Some people say that Yochanan "der hinkediker" (the lame) tailor did both. They say that he was one of the "lamed vov".[2] The following story is about Yochanan "der hinkediker" tailor's plan to instill dignity into the Jewish people of Zalleshitz during their last days in this world.

Like most enemies of the Jews, the Nazis perverted the knowledge they had of Judaism to further demoralize their victims. Knowing that the Sabbath was three days after Wednesday, the Nazi commander ordered the Jews of Zalleshitz on a Wednesday to gather in three days in the marketplace with all their belongings. It was rather unusual for the Nazis to choose Saturday for an "action" because they reserved weekends for rest and relaxation, but there were exceptions.

The Jews of Zalleshitz had been certain that the Nazis would be defeated before they reached their town. Now they were rudely awakened to the possibility of their annihilation. The leaders of the town called the people together in the synagogue to decide how to deal with this unexpected situation. The Rabbi of Zalleshitz was the first speaker. He spoke for hours suggesting different solutions to the problem but none seemed feasible. Then the president of the community spoke. His suggestions were equally impossible to implement. Suddenly, a Jew who was sitting in the rear of the synagogue rose to his feet and began screaming:

> "You don't know what to do, I'll tell you what to do!"

His name was Yochanan the tailor. Yochanan said:

> "Our fate has already been decreed, and we are powerless to change it. Therefore, I suggest that we spend the last three days of our lives in fervent prayer and praise of our G-d, so that we may die "al kiddush Hashem". Today is Wednesday and we will act as if tonight is "Yom Tov" (a holiday). We will start celebrating "Yom Tov" tonight and we will celebrate tomorrow and tomorrow night and Friday. We will conclude our celebration on the Sabbath."

The people were astonished by this suggestion, but after a while the logic of the suggestion took hold and a hum of excitement buzzed through the small synagogue. The people agreed that there was no better way to spend the last three days of life than by celebrating "Yom Tov". The Jews filed out of the synagogue, headed for their homes where they began to prepare for the festival. Together, families washed holiday clothes, baked challot, cooked special food and set festive tables. Towards evening, the Jews of Zalleshitz recited the "Kiddush" (sanctification of the festival wine) and proceeded to share their festival meal. From the corners of the town, holiday songs were audible. The fervor increased as Thursday morning approached and the Jews of Zalleshitz gathered in the synagogue for the morning prayers, interrupted only by the reading of the Torah. After the services, the Jews walked out of the synagogue. They were humming soft tunes, clasping hands and shuffling their feet to an invisible dancing rhythm.

Thursday night...Friday morning...the second day of the festival was nearly at an end when Yochanan the hinkediker tailor announced that the coming Sabbath would climax the joy of the past two days' celebration.

The Jews gathered early in the synagogue that Sabbath morning to be able to finish their prayers before the decreed time. At eleven o'clock, Yochanan the "hinkediker" tailor ascended the pulpit, opened the curtain of the holy ark and began reciting "Neila" (the prayer of the Day of Atonement recited just before the sun sets). The Jews of Zalleshitz solemnly joined him. At 11:45, Yochanan finished. He removed all the Torah scrolls from the ark, distributed them to the community leaders and called out to the Jews to follow him in prayer. They recited:

> "Shema Yisrael Adonay Elhaynu Adonay Echad" (Hear O Israel, the Lord our G-d, the Lord is One), (one time); "Boruch Shem Kevod Malchuto Leolam Vaed"[4] (Blessed Be His Holy Name For Ever and Ever), (three times); and "Adonay Hu Haelohim"[5] (G-d is The Righteous Judge), (seven times).

Then Yochanan, clutching a Torah scroll, descended the pulpit, motioned to the Jews of Zalleshitz to follow him, and together they walked to the marketplace still dancing and singing.

Thus the Jews of Zalleshitz, accepting the inevitable as a decree from the Almighty, marched to their death with dignity and pride.

1 Moshe Prager. Nitzotzay Gevura. Board of Jewish Education of New York, Inc. New York. 1966.

2 Talmud Bavli Succa 45 b. One of the thirty-six holy righteous men. It is believed that the world exists because of their meritorious lives.

3 Deuteronomy 6:4

4 Talmud Bavli Yoma 66 a

5 I Kings 18:39

YOSSEL RAKOVER'S APPEAL TO GOD[1]

In the ruins of the ghetto of Warsaw among heaps of charred rubbish, there was found packed tightly into a small bottle, the following testament, written during the ghetto's last hours by a Jew named Yossel Rakover.

"Warsaw, April 28, 1943.

"I, Yossel, son of David Rakover of Tarnopol, A Hasid of the Rebbe of Ger and a descendant of the great, pious and righteous families of Rakover and Meisel, inscribe these lines as the houses of the Warsaw ghetto go up in flames. The house I am in is one of the last unburnt houses remaining. For several hours an unusually heavy artillery barrage has been crashing down on us, and the walls around are disintegrating under the fire. It will not be long before the house I am in is transformed, like almost every other house of the ghetto, into a grave for its defenders. By the dagger-sharp, unusually crimson rays of the sun that strike through the small, half-walled-up window of my room through which we have been shooting at the enemy day and night, I see that it must now be late afternoon, just before sundown, and I cannot regret that this is the last sun that I shall see. All of our notions and emotions have been altered. Death, swift and immediate, seems to us a liberator, sundering our shackles; and beasts of the field in their freedom and gentleness seem to me to be so lovable and dear that I feel a deep pain whenever I hear the evil fiends that lord it over Europe referred to as beasts. It is untrue that the tyrant who rules Europe now has something of the beast in him. He is a typical child of modern man; mankind as a whole spawned him and reared him. He is merely the frankest expression of its innermost, most deeply buried instincts.

"In a forest where I once hid, I encountered a dog one night, sick and hungry, his tail between his legs. Both of us immediately felt the kinship of our situations. He cuddled up to me, buried his head in my lap, and licked my hands. I do not know if I ever cried so much as that night. I threw my arms around his neck, crying like a baby. If I say that I envied the animals at that moment, it would not be remarkable. But what I felt was more than envy. It was shame. I felt ashamed before the dog to be, not a dog, but a man. That is how matters stand. That is the spiritual level to which we have sunk. Life is a tragedy, death a savior; man a calamity, the beast an ideal; the day a horror, the night - relief.

"When my wife, my children and I - six in all - hid in the forest, it was the night and the night alone that concealed us in its bosom. The day turned us over to our persecutors and murderers. I remember with the most painful clarity the day when the Germans raked with a hail of fire the thousands of refugees on the highway from Grodno to Warsaw. As the sun rose, the airplanes zoomed over us. The whole day long, they murdered us. In this massacre, my wife with our seven-months-old child in her arms perished. And two others of my five remaining children also disappeared that day without a trace.

Their names were David and Yehuda; one was four years old, the other six.

"At sunset, the handful of survivors continued their journey in the direction of Warsaw, and I, with my three remaining children, started out to comb the fields and woods at the site of the massacre in search of the children. The entire night we called for them. Only echoes replied. I never saw my two children again, and later in a dream was told that they were in G-d's hands.

"My other three children died in the space of a single year in the Warsaw ghetto. Rachel, my daughter of ten, heard that it was possible to find scraps of bread in the public dump outside the ghetto walls. The ghetto was starving at the time, and the people who died of starvation lay in the streets like heaps of rags. The people of the ghetto were prepared to face any death but the death of hunger. Against no death did they struggle so fiercely as against death by starvation.

"My daughter, Rachel, told me nothing of her plan to steal out of the ghetto, an act punishable by death. She and a girl friend of the same age started out on the periolous journey. She left home under cover of darkness, and at sunrise she and her friend were caught outside the ghetto walls. Nazi ghetto guards, together with dozens of their Polish underlings, at once started in pursuit of these two Jewish children who had dared to venture out to hunt for a piece of bread in a garbage can. People witnessing the chase could not believe their eyes. One might have thought it was a pursuit of dangerous criminals, that horde of fiends running amok in pursuit of a pair of starved ten-year-old children. They did not endure very long in the unequal match. One of them, my child, running with her last ounce of strength, fell exhausted to the ground, and the Nazis then put a bullet through her head. The other child saved herself, but driven out of her mind, died two weeks later.

"The fifth child, Yaacov, a boy of thirteen, died on his Bar Mitzvah day of tuberculosis. The last child, my fifteen-year-old daughter, Chaya, perished during a Kinderaktion - children's operation - that began at sunrise last Rosh Hashona and ended at sundown. That day, before sunset, hundreds of Jewish families lost their children.

"Now my time has come. And like Job, I can say of myself, nor am I the only one that can say it - that I return to the soil naked, as naked as the day of my birth.

"I am forty-three years old, and when I look back on the past I can assert confidently, as confident as a man can be of himself, that I have lived a respectable, upstanding life, my heart full of love for G-d. I was once blessed with success, but never boasted of it. My possessions were extensive. My house was open to the needy. I served G-d enthusiastically, and my single request to Him was that He should allow me to worship Him with all my heart, and all my soul, and all my strength.

"I cannot say that my relationship to G-d has remained unchanged after everything I have lived through, but I can say with absolute certainty that my belief in Him has not changed by a hair's breadth. Previously, when I was well off, my relation to G-d was as to one who granted me a favor for nothing, and I was eternally obliged to Him for it. Now my relations to Him are as to one who owes me something, owes me much, and since I feel so, I believe that I have the right to demand it of Him. But I do not say like Job that G-d should point out my sin with His finger so that I may know why I deserve this; for greater and saintlier men than I are now firmly convinced that it is not a question of punishing sinners: something entirely different is taking place in the world. More exactly, it is a time when G-d has veiled His countenance from the world, sacrificing mankind to its wild instincts. This, however, does not mean that the pious members of my people should justify the edict, saying that G-d and His judgments are correct. For saying that we deserve the blows we have received is to malign ourselves, to desecrate the Holy Name of G-d's children. And those that desecrate our name desecrate the name of the Lord; G-d is maligned by our self-deprecation.

"In a situation like this, I naturally expect no miracles, nor do I ask Him, my Lord, to show me any mercy. May He treat me with the same indifference with which He treated millions of His people. I am no exception, and I expect no special treatment. I will no longer attempt to save myself, nor flee any more. I will facilitate the work of the fire by moisteninng my clothing with gasoline. I have three bottles of gasoline left after having emptied several scores over the heads of the murderers. It was one of the finest moments in my life when I did this, and I was shaken with laughter by it. I never dreamed that the death of people, even of enemies - even of such enemies - could cause me such great pleasure. Foolish humanists may say what they choose. Vengeance was and always will be the last means of waging battle and the greatest spiritual release of the oppressed. I had never until now understood the precise meaning of the expression in the Talmud that states:

> "Vengeance is sacred because it is mentioned between two of G-d's names." [2]:
>
> "El nekamot hashem".
> "A G-d of vengeance is the Lord."[3]

"I understand it now. I know now why my heart is so overjoyed at remembering that for thousands of years we have been calling our Lord a G-d of Vengeance: A G-d of Vengeance is our Lord. We have had only a few opportunities to witness true vengeance. When we did however, it was so good, so worthwhile - I felt such profound happiness, so terribly fortunate - that it seemed an entirely new life was springing up in me. A tank had suddenly broken into our street. It was bombarded with flaming bottles of gasoline from all the embattled houses. They failed to hit their target, however, and the tank continued to approach. My friends and I waited until the tank was almost upon us. Then, through the half bricked-up window, we suddenly attacked. The tank soon burst into flames, and six blazing Nazis jumped out. Ah, how they burned! They burned like the Jews they had set on fire. But they shrieked more. Jews do not shriek. They welcome death like a savior. The Warsaw ghetto perished in battle. It went down shooting, struggling, blazing, but not shrieking!

"I have three more bottles of gasoline. They are as precious to me as wine to a drunkard. After pouring one over my clothes, I will place the paper on which I write these lines in the empty bottle and hide it among the bricks filling the window of this room. If anyone ever finds it and reads it, he will, perhaps, understand the emotions of a Jew, one of millions, who died forsaken by the G-d in whom he believed unshakeably. I will let the two other bottles explode on the heads of the murderers when my last moment comes.

"There were twelve of us in this room at the outbreak of the rebellion. For nine days we battled against the enemy. All eleven of my comrades have fallen, dying silently in battle, including the small boy of about five - who came here only G-d knows how and who now lies dead near me, with his face wearing the kind of smile that appears on children's faces when dreaming peacefully - even this child died with the same epic calm as his older comrades. It happened early this morning. Most of us were dead already. The boy scaled the heap of corpses to catch a glimpse of the outside world through the window. He stood beside me in that position for several minutes. Suddenly he fell backwards, rolling down the pile of corpses, and lay like a stone. On his small, pale forehead, between the locks of black hair, there was a spattering of blood.

"Up until yesterday morning when the enemy launched a concentrated barrage against this stronghold, one of the last in the ghetto, at sunrise every one of us was still alive although five were wounded. During yesterday and today, all of them fell, one after the other, one on top of the other, watching and firing until shot to death. I have no more ammunition, apart from the three bottles of gasoline. From the floors of the house above still come frequent shots, but they can hold out no more hope for me, for by all signs, the stairway has been razed by shell fire and I think the house is about to collapse. I write these lines lying on the floor. Around me lie my dead comrades. I look into their faces, and it seems to me that a quiet but mocking irony animates them as if they were saying to me: 'A little patience, you foolish man, another few minutes and everything will become clear to you too.' This irony is particularly noticeable on the face of the small boy lying near my right hand as if he were asleep. His small mouth is drawn into a smile exactly as if he were laughing, and I, who still live and feel and think - it seems to me that he is laughing at me. He laughs with that quiet but eloquent, penetrating laughter so characteristic of the wise, speaking of knowledge with the ignorant who believe they know everything. Yes, he is omniscient now. Everything is clear to the boy now. He even knows why he was born, but had to die so soon, why he died only five years after his birth. And even if he does not know why, he knows at least that it is entirely unimportant and insignificant whether or not he knows it, in the light of the revelation of that godly majesty of the better world he now inhabits, in the arms of his murdered parents to whom he has returned.

In an hour or two, I will make the same discovery. Unless my face is eaten by the flames, a similar smile may also rest on it after my death. Meanwhile, I still live, and before my death I wish to speak to my Lord as a living man, a simple, living person who had the great but tragic honor of being a Jew.

"I am proud that I am a Jew not in spite of the world's treatment of us, but precisely because of this treatment. I should be ashamed to belong to the people who spawned and raised the criminals who are responsible for the deeds that have been perpetrated against us.

"I am proud to be a Jew because it is an art to be a Jew, because it is difficult to be a Jew. It is no art to be an Englishman, an American or a Frenchman. It may be easier, more comfortable, to be one of them, but not more honorable. Yes, it is an honor to be a Jew.

"I believe that to be a Jew means to be a fighter, an everlasting swimmer against the turbulent, criminal human current. The Jew is a hero, a martyr, a saint. You, our enemies, declare that we are bad? I believe that we are better and finer than you, but even if we were worse - I should like to see how you would look in our place!

"I am happy to belong to the unhappiest of all peoples of the world whose precepts represent the loftiest and most beautiful of all morality and laws. These immortal precepts which we possess have now been even more sanctified and immortalized by the fact that they have been so debased and insulted by the enemies of the Lord.

"I believe that to be a Jew is an inborn trait. One is born a Jew exactly as one is born an artist. It is impossible to be released from being a Jew. That is our godly attribute that has made us a chosen people. Those who do not understand this will never understand the higher meaning of our martyrdom. If I ever doubted that G-d once designated us as the chosen people, I would believe now that our tribulations have made us the chosen one.

"I believe in You, G-d of Israel, even though You have done everything to stop me from believing in You. I believe in Your laws even if I cannot excuse Your actions. My relationship to You is not the relationship of a slave to his master but rather that of a pupil to his teacher. I bow my head before Your greatness, but will not kiss the lash with which You strike me.

"You say, perhaps, that we have sinned, O Lord? It must surely be true. And therefore we are punished? I can understand that too. But I should like You to tell me - Is there any sin in the world deserving of such punishment as the punishment we have received?

"You assert that You will yet repay our enemies? I am convinced of it. Repay them without mercy? I have no doubt of that either. I should like You to tell me, however - Is there any punishment in the world capable of compensating for the crimes that have been committed against us?

"You say, I know, that it is no longer a question of sin and punishment, but rather a situation in which Your countenance is veiled, in which humanity is abandoned to its evil instincts. I should like to ask You, O Lord - and this question burns in me like a consuming fire - What more, O, what more must transpire before You unveil Your countenance again to the world?

"I want to say to You that now, more than in any previous period of our eternal path of agony, we, we the tortured, humiliated, buried alive, and burned alive, we the insulted, the object of mockery, we who have been murdered by the millions, we have the right to know: 'What are the limits of Your forebearance?'

"I should like to say something more: Do not put the rope under too much strain, lest, alas, it may snap. The test to which You have put us is so severe, so unbearably severe, that You should - You must - forgive members of Your people, who in their misery have turned from You.

"Forgive those who have turned from You in their misery, but also those who have turned from You in their happiness. You have transformed our life into such a frightful, perpetual struggle that the cowards among us have been forced to flee from it; and what is happiness but a place of refuge for cowards? Do not chastise them for it. One does not strike cowards, but has mercy on them, rather than us, O ' Lord. Have mercy on them, rather than us, O Lord.

"Forgive those who have desecrated Your name, who have gone over to the service of other gods, who have become indifferent to You. You have castigated them so severely that they no longer believe that You are their Father, that they have any Father at all.

"I tell You this because I do believe in You, believe in You more strongly than ever, because now I know that You are my Lord, because after all You are not; You cannot after all be the G-d of those whose deeds are the most horrible expression of ungodliness.

"If you are not my Lord, then whose Lord are You? The Lord of the murderers?

"If those that hate me and murder me are so benighted, so evil, what then am I if not the person who reflects something of Your light, of Your goodness?

"I cannot extol You for the deeds that You tolerate. I bless You and extol You, however, for the very fact of Your existence, for Your awesome mightiness.

"The murderers themselves have already passed sentence on themselves and will never escape it; but may You carry out a doubly severe sentence on those who are condoning the murder.

"Those that condemn murder orally, but rejoice at it in their hearts ... Those who meditate in their foul hearts: It is fitting, after all, to say that he is evil, this tyrant, but he carries out a bit of work for us for which we will always be grateful to him!

"It is written in your Torah that a thief should be punished more severely than a brigand, in spite of the fact that a thief does not attack his victim physically and merely attempts to take his possessions stealthily.

"The reason for this is that the brigand in attacking his victim in broad daylight, shows no more fear of man than of G-d. The thief on the other hand fears man, but not G-d. His punishment, therefore, is greater.

"I should be satisfied if you dealt with the murderers as with brigands, for their attitude towards you and towards us is the same.

"But those who are silent in the face of murder, those who have no fear of you, but fear what people might say (fools! they are unaware that the people will say nothing!), those who express their sympathy with the drowning man but refuse to rescue him - punish them, O Lord, punish them, I implore You, like the thief, with a doubly-severe sentence!

"Death can wait no longer. From the floors above me, the firing becomes weaker by the minute. The last defenders of this stronghold are now falling, and with them falls and perishes the great, beautiful and G-d-fearing Jewish part of Warsaw. The sun is about to set, and I thank G-d that I will never see it again. Fire lights the small window, and the bit of sky that I can see is flooded with red like a waterfall of blood. In about an hour at the most I will be with the rest of my family and with the millions of other stricken members of my people in that better world where there are no more doubts.

"I die peacefully, but not complacently; persecuted, but not enslaved; embittered, but not cynical; a believer, but not a supplicant; a lover of G-d, but no blind amen-sayer of His.

"I have followed Him even when He repulsed me. I have followed His commandments even when He castigated me for it; I have loved Him and I love Him even when He has hurled me to the earth, tortured me to death, made me an object of shame and ridicule.

"My Rabbi would frequently tell the story of a Jew who fled from the Spanish Inquisition with his wife and child, striking out in a small boat over the stormy sea until he reached a rocky island. A flash of lightning killed his wife; a storm rose and hurled his son into the sea. Then, as lonely as a stone, naked, barefoot, lashed by the storm and terrified by the thunder and lightning, hands turned up to G-d, the Jew, again setting out on his journey through the wastes of the rocky island, turned to G-d with the following words:

"G-d of Israel, I have fled to this place in order to worship You without molestation, to obey Your commandments and sanctify Your name. You, however, have done everything to make me stop believing in You. Now, lest it seem to You that You will succeed by these tribulations in driving me from the right path, I notify You, my G-d and the G-d of my father, that it will not avail you in the least. You may insult me, You may castigate me, You may take from me all that I cherish and hold dear in the world, You may torture me to death - I will believe in You, I will always love You!

"And these are my last words to You, my wrathful G-d: nothing will avail You in the least. You have done everything to make me renounce You, to make me lose my faith in You, but I die exactly as I have lived, crying:

"Eternally praised be the G-d of the dead, the G-d of vengeance, of truth and of law, Who will soon show His face to the world again and shake its foundations with His almighty voice.

> 'Sh'ma Yisrael Adonay Elohaynu Adonay Echad.'[4]
> 'Hear O Israel, the Lord our G-d the Lord is One.'
>
> 'B'yadcha afkeed ruchi.'[5]
> 'Into your hands, O Lord, I consign my soul.'"

1 Zvi Kolitz. The Tiger Beneath The Skin, "Yossel Rakover's Appeal to G-d". Farrar, Straus, & Giroux, Inc. New York. This story was written by Zvi Kolitz as a memorial to the victims of the Holocaust.

2 Talmud Bavli Berachot 33 a

3 Psalms 94:1

4 Deuteronomy 6:4

5 Psalms 31:6

REDEMPTION OF CAPTIVE JEWS

פדיון שבויים

CHAPTER THREE

"The duty of redeeming captives supersedes the duty of charity."[1]

A Mitzva of paramount importance in Jewish life is "Pidyon Shevuyim" (the redemption of Jews captured by slave dealers, pirates, warriors, or imprisoned unjustly by the authorities).[1]

The mitzva of redeeming captives supersedes the mitzva of giving charity because it develops compassion, mercy, and solidarity among Jews who are responsible to care for each other. A Jew is not allowed to close his eyes from the suffering of his captured brethren because he transgresses many commandments concerning his relationship to his fellow Jew, most important of which is: "Love your neighbor as you love yourself."[2] This mitzva is so important that charity money collected for the upkeep of a synagogue may be used to redeem captives.

The redeemer is considered as one who has fulfilled a precept of the Torah because he saves men from torture and women from dishonor, especially since the captors have the opportunity to inflict upon the captives whatever suffering they wish.

The mitzva of "pidyon shevuyim" must be performed as quickly as possible since procrastinators are compared to murderers.[3]

> "To perish by the sword is worse than a natural death, to die of hunger worse still, and captivity worst of all."[4]

Unfortunately, Jews have had many occasions to fulfill this mitzva throughout Jewish history. The first Jew to fulfill the mitzva of "pidyon shevuyim" was our forefather Abraham:

> "And when Abram heard that his kinsman had been taken prisoner, he led out all his trusted followers, born in his own household, three hundred and eighteen, and went in pursuit as far as Dan...And he recovered all the possessions, and also returned his kinsman with his possessions, as well as the women and the people."[5]

During the Middle Ages, Jews who resided in western European countries were accused of blood libels and were imprisoned. Often nobles imprisoned Jews for no other reason than to extort large sums of money from the Jewish community for the enrichment of their private treasuries. These nobles knew that Jews would go to great expense to redeem their captured brethren.

Throughout the horror of the Chmielnicki pogroms in the middle of the seventeenth century, special committees were appointed by the "Vaad Arba Aratzot" (the Council of the Four Lands)[6] to redeem captured Jews from Tatar and Cossack soldiers who roamed the Ukraine and Volhynia, Poland, and kidnapped Jews at will. The committees were encouraged to act quickly by the "Vaad Arba Aratzot", and a "minyan" of ten adult males in any village that found out about a captured Jew was empowered to take money from the general budget in order to prevent the delay of fulfilling this mitzva. Every Jewish village, under the jurisdiction of the "Vaad Arba Aratzot" had redemption money set aside for this purpose.

This mitzva is still of paramount importance today since so many of our brethren are imprisoned unjustly behind the Iron Curtain and in some Arab countries.

The following stories describe the performance of "pidyon shevuyim" and its effect on the redeemed captives.

1 Talmud Bavli Baba Batra 8 b

2 Leviticus 19:18

3 Maimonides. Matnat Evyonim 14

4 Talmud Bavli Baba Batra 8 b

5 Genesis 14:14,16

6 The Council of the Four Lands was made up of both laymen and rabbinical scholars who represented the four sections of Poland. It directed manifold social, ethical and legal matters which concerned the Polish Jewish community.

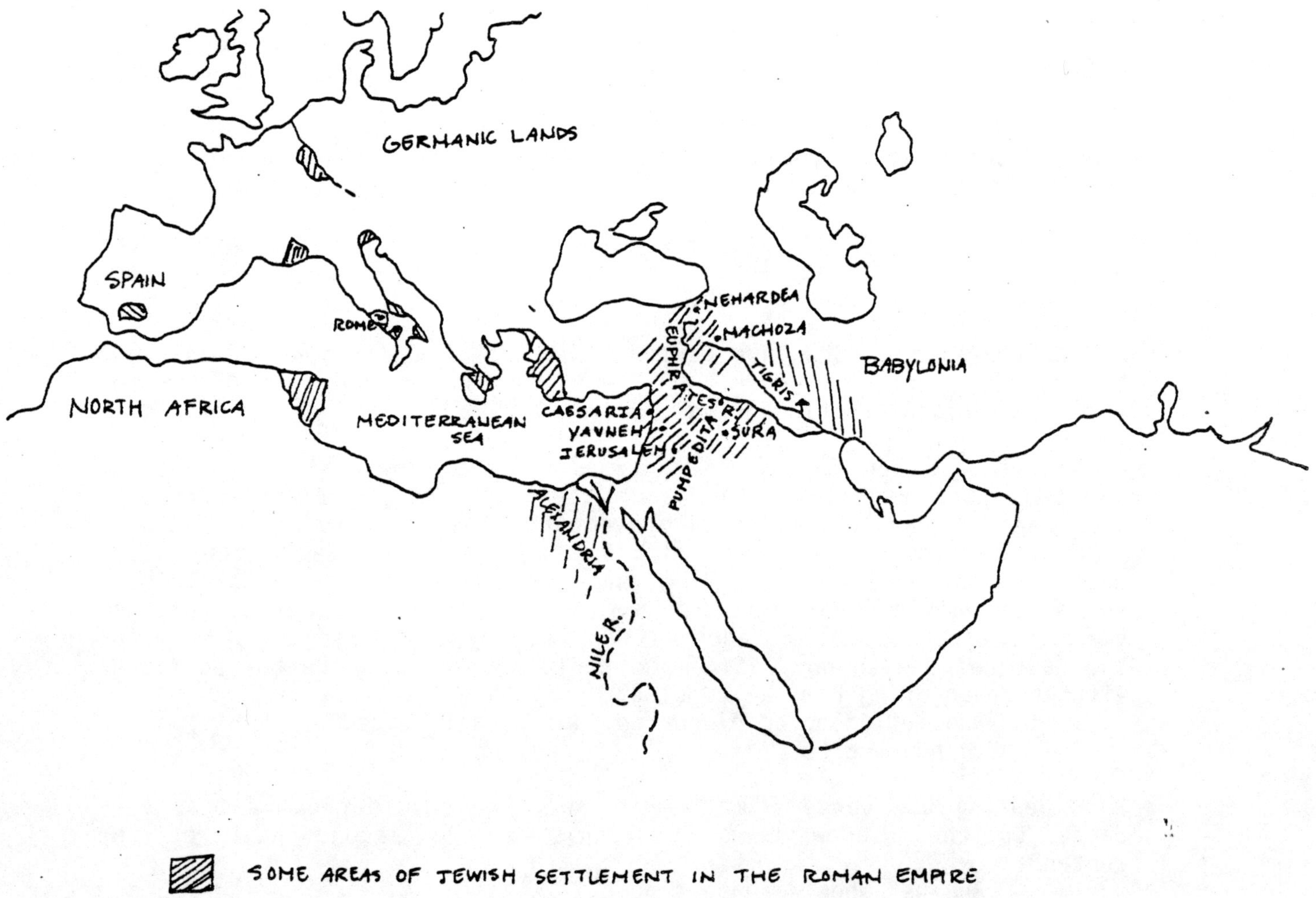
GERMANIC LANDS
SPAIN
ROME
NORTH AFRICA
MEDITERRANEAN SEA
CAESARIA
YAVNEH
JERUSALEM
ALEXANDRIA
NILE R.
NEHARDEA
MACHOZA
EUPHRATES R.
PUMPEDITA
SURA
TIGRIS R.
BABYLONIA
SOME AREAS OF JEWISH SETTLEMENT IN THE ROMAN EMPIRE

THE CAPTIVE BOY[1]

During the years of the Roman conquest of Eretz Yisrael, many Jews were led into captivity. Some were sold as slaves while others were fed to the lions of the Roman circuses. Those Jews who were sold as slaves hoped for redemption at the hands of their brethren who had become traders and settled in Roman ruled lands prior to the conquest of Eretz Yisrael.

Rabbi Yehoshua ben Chananya was one of the outstanding leaders in the generation following the destruction of the second Holy Temple, around the beginning of the second century of the common era. He was respected as a spokesman for the Jewish people, and he travelled occasionally to Rome on missions of mercy for the Jewish community in Eretz Yisrael. On one trip, the Jews residing in Rome told him that among the newly arrived captives brought by the Roman soldiers from Eretz Yisrael was a young, handsome, intelligent looking Jewish boy. The informers did not know what the Romans planned to do with him.

Rabbi Yehoshua decided to investigate the matter. If the situation proved to be as the informers described, he knew he would do everything in his power to free the youngster.

Rabbi Yehoshua ben Chananya went to the prison and tried to question the guards about the young boy. The guards did not permit him to enter the prison or talk to the captive boy.

Rabbi Yehoshua ben Chananya was not dissuaded from his determination to find out more information about the boy. He thought awhile, and then decided upon a plan to circumvent the guard's refusal to let him enter the prison. He began walking around the outside of the jail, muttering to himself, pretending to be immersed in his own troubles. As he was walking, he noticed a small, pale faced boy peering through the window bars. Rabbi Yehoshua ben Chananya wondered if this were the captured Jewish boy. Since he could not talk to the boy outright without endangering him, he mumbled:

> "Who let Jacob be plundered, and Israel robbed?
> Did not the Lord?"

Upon hearing the verse from the prophet, the boy placed his head very close to the window bars and finished the second half of the sentence:

> "Against Whom we have sinned,
> And in Whose Ways we were not willing to walk
> And Whose Laws we did not obey."[2]

Rabbi Yehoshua ben Chananya realized that the captive boy was indeed the Jewish prisoner. Pained at his suffering, he tearfully swore that he would not leave Rome until he had redeemed him, no matter the amount of the ransom required.

Rabbi Yehoshua ben Chananya paid a heavy ransom to the Roman government for the release of the young Jewish boy. Both of them returned to Eretz Yisrael. The boy immersed himself in the study of Torah and became a great scholar. His name was Rabbi Yishmael ben Elisha.[3]

1 Talmud Bavli Gitin 58 a
2 Isaiah 42:24
3 Rabbi Yishmael ben Elisha Kohen Gadol was a "Tanna" (the group of Rabbis who composed the Mishna). He expanded to 13, the number of hermeneutical principles by which the Torah is expounded. He was one of the "Asarah Harugay Malchut" (10 illustrious sages who died to sanctify G-d's Holy Name).

RANSOMED RABBIS SPREAD TORAH THROUGHOUT THE MOSLEM DIASPORA[1]

During the Middle Ages, Jewish traders who lived in Moslem countries feared being captured by pirates while they travelled for business on the high seas.

Alexandria, Egypt, was one of the leading Jewish communities in the Moslem lands. Because of its geographical location, the Jews of Alexandria were fully aware of their responsibility to redeem captives that would be auctioned in their port, and they created a special fund, endowed by the wealthier members of the community which was earmarked for this specific purpose.

Four leading Rabbis set sail from Bari in southern Italy to attend an intensive Torah seminar and to collect money on behalf of the Babylonian academies to be used as dowries for poor brides.

They were captured by pirates who knew the great profit four distinguished scholars could bring them. They were held captive until they reached four major ports along the Mediterranean coast.

Since the Jews in the Diaspora were accustomed to redeeming their captured brethren; each community redeemed one of the four Rabbis. Rabbi Shemariah ben Rabbi Elchanan[2] was auctioned as a slave in the large Jewish community of Alexandria, Egypt; Rabbi Chushiel ben Rabbi Elchanan[3] was auctioned in Kairouan, Tunisia; Rabbi Moses ben Chanoch[4] in Cordova, Spain; and the fourth may have been Rabbi Natan HaBavli in Narbonne, France.

When the redeemers realized that these captives were distinguished scholars, they asked them to remain and assume leadership in those communities that had paid for their redemption. These acts of "pidyon shevuyim" resulted in the establishment of new schools of Talmudic learning, spreading the word of G-d throughout the countries of the Moslem Diaspora.

Rabbi Moses ben Chanoch, the redeemed captive who settled in Cordova, Spain, was so humble that he did not disclose his identity. No one knew of his great scholarship until he volunteered the answer to a legal problem which even the head of the reknowned Cordova Talmudic Academy could not resolve. The community immediately recognized this humble man and appointed him to the leadership of the Cordova Talmudic Academy.

1 Abraham ben David Halayvi ibn Daud. The Book of Tradition: An Authoritative History of Spanish Jewry.

2 Rabbi Shemariah ben Elchanan became the head of the Talmudic Academy in Fostat (Cairo). He wrote a commentary on the Song of Songs.

3 Rabbi Chushiel ben Rabbi Elchanan was the leader of the Talmudic cademy in Kairouan, Tunisia and it became the center for Torah study in North Africa.

4 Rabbi Moses ben Chanoch was head of the Yeshiva in Cordova, Spain. He responded to many halachic (legal) questions that were sent to him from other Jewish communities. His leadership marked the beginning of independent scholarship of the Spanish Jewish community. It was about this time that the Babylonian Jewish community began its decline.

5 Very little is known about the fourth captive's settlement in Narbonne. There was a tenth century scholar of the same name who was one of the leaders of the Babylonian Jewish community.

RABBI MAYER REFUSES TO BE RANSOMED[1]

Throughout the latter half of the Middle Ages, from about the twelfth century, the Jews of western Europe lived in fear and were in constant danger of being kidnapped for ransom. Clergy or nobles perpetrated plots against the Jews living under their protection in order to squeeze ransom money from the Jewish community, for they knew that the Jewish community would try their utmost to ransom those captured. The ransom 'money' was usually used to fill the coffers of the individual priest or noble to cover the expenses of their extravagant living habits.

The Jewish communities in western Europe were pathetically poor. However, a few individual Jews in each community were fortunate to rise above the poverty level and were willing to share their wealth to redeem captives when necessary.

This story is about Rabbi Mayer who lived in Rothenburg, one of the towns in the Germanic lands of western Europe around the thirteenth century. He was imprisoned in order that the treasury of Emperor Rudolph I of Hapsburg be enriched. However, Rabbi Mayer refused to permit the Jewish community to ransom him because he feared he would set a dangerous precedent that might open the floodgates of increased kidnappings of other prominent Jews.

Rabbi Mayer was respected by the learned throughout western Europe for his scholarly understanding of Torah and its application to daily life. His home was always open to those in need, and people came from all over the Germanic lands seeking his advice.

The Count of Rothenburg cultivated the friendship of Rabbi Mayer, and the two of them spent many hours together in intellectual discussion. Benefiting from this friendship were the Jews of Rothenburg who lived in relative safety.

Rabbi Mayer prided himself on the achievement of his students. All of them, with the exception of one named Kempe, were serious scholars. Kempe was mischievous and cunning. When Rabbi Mayer became aware of the potential harm which Kempe could cause, he expelled him from his "Yeshiva". Kempe retaliated by becoming an informer and threatened the welfare of the entire Jewish community.

Kempe convinced the Count that the Jewish community had "hidden" wealth which could be used as an additional source of income for Emperor Rudolph I of Hapsburg's treasury. The Count decided to extract this "hidden" money from the Jewish community in order to find favor in the eyes of the Emperor. The next time that the Count and Rabbi Mayer met for intellectual discussion, the Count demanded of Rabbi Mayer that he raise an exorbitant sum of money in two short months or face life imprisonment. The Count was confident that the Jewish community would certainly pay the required sum to prevent the jailing of Rabbi Mayer.

When Rabbi Mayer returned home, he began to write letters to the heads of other Jewish communities informing them of the plight of the Jews of Rothenburg, calling their attention to the emergency situation while beseeching their help. He also made preparations for an alternate plan should the replies to his letters be negative. The alternate plan was for him to flee Germany and sail for Eretz Yisrael. There he hoped to resettle and spend the remaining years of his life, dedicated to teaching Torah.

Rabbi Mayer's wife and children had been murdered many years before in a pogrom. Only one daughter, Sarah, escaped with her life. Since she was of marriageable age, Rabbi Mayer decided to marry her off to his best student, Asher, whom he loved as his own son. He wanted all the arrangements to be secret so that the young couple would be able to escape from Rothenburg to start a new life together in Spain, a country that welcomed Jews at that time.

As he suspected, Rabbi Mayer received very few answers to his letters. Most of them offered sympathy but no concrete help because the majority of the Jewish communities in the Germanic lands were pitifully poor. He decided to implement his alternate plan to arrange the wedding and flee immediately afterward.

Rabbi Mayer decided that Asher and Sarah should travel separately to lessen suspicion that he was fleeing. He wanted people to believe that he was simply travelling to visit another Jewish community. He arranged to meet his children in Goerz, Lombardy, so they could say goodbye before each set out for their respective destinations. As he was waiting for his children in Goerz while resting from his arduous journey, he was recognized by Kempe who had moved to that city.

Kempe informed Count Meinharat of Goerz that the aging Rabbi had fled to escape paying the huge sum of money which was demanded of the Jewish community of Rothenburg. Count Meinharat arrested Rabbi Mayer and delivered him to Rudolph I.

The Emperor imprisoned Rabbi Mayer in Wasserburg and demanded the fantastic ransom of 23,000 pounds of silver for his release. After a while, Rabbi Mayer was transferred to a prison at Ensisheim in the province of Alsace.

When Asher and Sarah arrived in Goerz, they were horrified to discover that her father had been imprisoned. They immediately set out for the prison in Ensisheim. There they found that Rabbi Mayer was allowed visitors and he told Sarah and Asher the conditions of his release. When they offered to raise the ransom money, he forbade them to do it. He explained that were he to allow them to raise the money, it would establish a precedent; that every Rabbi would be threatened with the same imprisonment in order to support the Emperor. He pleaded with Asher and Sarah to leave for Spain before the Emperor would also imprison them.

Rabbi Mayer remained imprisoned by the Emperor for seven years until the day of his death. During the years of Rabbi Mayer's imprisonment, the Emperor made one concession. He permitted Rabbi Mayer's library of holy books to be brought to the prison. Rabbi Mayer spent his days spiritually free but physically enslaved. When he died, he was buried in the prison grounds because the Emperor refused to release his body without the payment of the ransom.

Fourteen years after his death, a great philanthropist named Alexander Wimpfen arranged for the release of his body by the payment of a gigantic sum of ransom money to the Emperor's treasury. The conditions of the ransom stipulated that the Emperor receive the money for his treasury and Alexander Wimpfen be guaranteed burial alongside Rabbi Mayer in the Jewish cemetery of Worms.

1 Marcus Lehman. Five Novellettes. Merkos L'Inyonay Chinuch Inc. Brooklyn, New York. 1964. This story was adapted.

THE REWARD[1]

Around the middle of the seventeenth century, chaos erupted and disturbed the peaceful sojourn of the Jews who had settled in Poland. The confusion was caused by the utter hopelessness of the Jews after they had been deluded into believing that Shabbtai Zevi was the Messiah and by their feeling of helplessness after the Bogdan Chmielnicki massacres. The Jews who survived the Chmielnicki pogroms became very poor.

To add to their misery, Austria, Prussia, and Russia thrice partitioned Poland during the eighteenth century. The area which Russia acquired incorporated a section of eastern Europe that was home to half of the world's Jewish population. This enlarged area of Jewish habitation was called the Pale of Settlement and it was bounded on the north and south by the Baltic and Black Seas, and on the east and west by the partitioning countries.

The Jews who lived in the Pale of Settlement lived in "shtetlach" (plural of shtetl), villages predominantly inhabited by Jews, far removed from the large cities. They eked out a living working in agriculture, textiles, crafts and manufacturing iron work. Some Jews leased inns from the local "poretz" (the noble land owner) and other Jews leased liquor stores, supplying the neighborhood tenant farmers with whiskey.

Many of the "preetzim" (plural of poretz) were tyrants who made unreasonable demands on the people that leased their lands. They extracted exorbitant rents and even greater taxes that were increased each year to enrich their treasuries.

The downtrodden Jews did no know how to cope with the cruel demands of the "preetzim". They were afflicted with the misery of eking out a living. When Rabbi Yisrael ben Eliezer, the Baal Shem Tov, the founder of the Chassidic movement, appeared on the scene, he taught the Jews to convert "Tzarah" (trouble) into "Tzohar" (radiant light).

This story shows how the Baal Shem Tov brought radiant light and joy to a young couple that had ransomed a tenant farmer and his family.

A Jewish man had leased a liquor store from a cruel "poretz" who was the master over a certain Polish village. Unfortunately, business in the liquor store was not very lucrative and the storekeeper did not earn enough profit to pay the rent. The Jew did not despair. Rather, he prayed that G-d would bless him with better fortune the following year, so that he would be able to pay both the back rent and the rent for the coming year. Then he approached the "poretz" with the proposal to wait until the following year when he would be able to clear his debt completely. Surprisingly, the "poretz" agreed to wait for his money. But the next year came and went, and the storekeeper's financial situation remained unchanged. His debt to the "poretz" amounted to 300 rubles.

By this time, the "poretz" had exhausted his patience with the delinquent storekeeper, so he decided to jail him, his wife and children. He ordered the jail guards to throw the imprisoned family a crust of bread and some water each day in order to sustain them.

After a while, the "poretz" realized that keeping the storekeeper imprisoned was not a solution to the problem of collecting the money which he owed. The "poretz" thought a long time and concluded that the most feasible plan would be to take the unfortunate storekeeper and his family to the market place and offer them as captive slaves to be ransomed by the Jewish community. The "poretz" realized that if it were known that he planned to kill this family unless a 300 ruble ransom were paid, some merciful Jew would certainly pay for their redemption. This way, he reasoned, he was certain to collect the rent that the storekeeper owed. The plan seemed plausible.

On the next market day, an officer of the "poretz" shackled the storekeeper and his family, hoisted them into a wagon and set out for the market place. The officer was instructed to announce publicly that if the family were not redeemed for 300 rubles, they would be sentenced to death.

Many Jews passed the site where the storekeeper and his family languished, shackled in the rear of the wagon. They looked with compassion upon the unfortunate people, but were powerless to help because none of them had the required sum of money.

One passerby was a young man who was filled with anguish at the plight of the unfortunate family. He was employed as a servant in various Jewish homes and he had accumulated a savings of 150 rubles. He felt badly that he did not have the required sum. He stood by to wait and see what would happen.

While he waited, a young orphan girl who was employed as a maid for a Jewish family passed the shackled family. Her soul was also troubled by their horrible plight. When she saw the young man standing near the wagon, she began to discuss the captive family's plight with him. She volunteered the information that she had saved 150 rubles that she would be willing to use for their redemption money. The young man realized that the sum of their savings would provide the necessary amount to redeem the family.

Until that moment, the unfortunate family was certain that the "poretz" would carry out his evil intent. Their joy was without bounds when the "poretz's" officer released them from their shackles after he had received the ransom money. They thanked G-d for performing miracles. Then they returned to their home in the village.

The "mitzva" which the two young people performed made a tremendous impression upon the heavenly court. It decided that these two people who had sacrificed so much for the performance of a "mitzva" would be fitting soulmates. It was revealed to the Baal Shem Tov in a dream that he would have the privilege of arranging this wedding.

The Baal Shem took two of his best students and set out for the village where the young couple lived. In his possession were bridal garments. When he arrived at his destination, he arranged lodging for himself in an inn. He also instructed the innkeeper to begin preparations for a wedding feast.

The next day the Baal Shem sent his students to fetch the young man. When he arrived, the Baal Shem proposed that he marry the young girl who had shared with him the "mitzva" of redeeming the captives. The young man was agreeable. Then the Baal Shem sent for the young girl, and in answer to the Baal Shem's question, she agreed to marry the young man. The Baal Shem notified the innkeeper that the wedding would take place in seven days.

On the day of the wedding, all the villagers gathered in the inn to witness the marriage of the couple who had shared the "mitzva" of redeeming the captives. At the wedding feast, the Baal Shem called upon the leaders of the village to bless the young couple. After the leaders finished blessing the newlyweds, three strangers who were seated in the back of the dining room stood up. They were dressed in fine garments, and their faces had a special sparkle. The first one said:

> "I bless you with orchards."

The second one said:

> "I bless you with an inn that will be your source of sustenance."

The third one said:

> "I bless you with precious jewels."

Afterwards, the wedding guests departed.

The next morning, the groom awoke to the realization that he had the responsibility of earning a livelihood. He had a wealthy uncle in a distant village, so he decided to borrow a horse and wagon from a friend, and rode to visit his uncle. He wanted to ask his uncle for a loan of 100 rubles to start a small business.

The friend lent the groom a horse and wagon and he started off for his uncle's village. He had to travel through open fields and cross bridges that spanned flowing streams. He was completely alone with the silence of nature.

Suddenly he heard a plea for help coming from beneath a bridge that he was crossing. He jumped off the carriage and ran to the bank of the stream. Struggling to reach the bank was a little boy, almost completely exhausted from attempting to reach land. The groom jumped into the stream, grabbed the boy, and swam with him to safety. He put the boy in back of the wagon, covered him with a wagon mat, and asked him where he lived. The boy said he was the son of a nearby "poretz" and directed the groom to his home.

The boy said:

> "I was walking in the fields last evening, and I didn't realize how rapidly night was approaching. I could not find my way back to my home in the dark. I was tired and frightened. Early this morning, I started out for home. I slipped from the bridge and fell into the stream, but I could not swim to shore. Thank you for saving my life."

The groom pointed the horses in the direction of the boy's house. When they reached the "poretz's" house, the groom carried the boy inside. The "poretz" and his wife were overjoyed to see their only son. They had sent a search party out after he had failed to return home, but the searchers were unsuccessful. The parents feared they would never see their son again.

They sat with their son in the living room and comforted him. They changed his clothes and brought him food and drink. They completely ignored the groom who was standing shyly in the vestibule. When the boy was warm and comfortable, he realized that the groom was still standing in the vestibule. He pointed this out to his parents. The "poretz" insisted that the groom join the family and said to him:

> "I must reward you for saving my son's life. I will give you the inn in the village and the orchards surrounding the inn."

The "poretz's" wife excused herself for a moment, ran to her room, grabbed her jewelry box and returned to the living room. She gave the jewelry box to the groom.

The groom accepted the gifts, bid farewell to the "poretz", his wife and their son, and started out for his village. As he was travelling, he realized that the blessings of the three special wedding guests had been fulfilled.

1 M.S. Kasher and Y.Y. Belkrovitz. Anshay Maaleh. Maarav Publishing Co. Jerusalem. 1974. This story was adapted.

MY REBBE AND THE UKRAINIAN THIEF[1]

There are approximately three million Jews behind the Iron Curtain. They are the children and the grandchildren of those Jews who lived in the Pale of Settlement, the area bridging eastern and western Europe, which Russia annexed in various wars during the past two centuries. Their civil liberties are restricted and they are in danger of complete spiritual annihilation. Their voices cry out to freedom lovers in the west for release from bondage.

This story is about the miraculous redemption of a Jewish captive. The news of his plight somehow leaked to the free world. Protests and petitions, the two modern methods which help in "pidyon shevuyim", were used to secure his release from a Siberian jail. He flew to Eretz Yisrael and settled in the city of Benay Brak where he lives with his wife, children and grandchildren. This is his story.

"I was incarcerated in a Siberian jail, together with my Rebbe. We had to share the cell with a Ukrainian thief. It was Passover night, the night of the first Sayder, and my Rebbe turned to me and said:

> 'Tonight we have neither matza nor wine to make a Sayder, but we can repeat the story of the Exodus from Egypt.'

"He led me to the corner of the cell, opposite the thief, and we huddled together. My Rebbe began chanting the Haggadah, the story of the Exodus:

> 'We were slaves to Pharaoh in Egypt and if Almighty G-d had not redeemed us, we, and our children, and our children's children would still be slaves.'

"The thief called out from his corner:

> 'Keep quiet. You are disturbing me. If you don't stop singing Jewish stories, I will call the guard.'

"My Rebbe paid no attention to the thief's complaints and continued his recitation of the Haggadah. The thief, true to his word, began screaming and hollering. The guard, attracted by the commotion, cam running.

"The guard opened the cell door, leered at my Rebbe and said angrily:

> 'If I hear any more commotion, I will return and shoot all of you.'

Then he slammed the door, and stalked away.

"My Rebbe turned to the Ukrainian thief and said softly:

> 'Don't you have any self respect? The Russians have invaded your land. They are destroying your people. Aren't you emotionally disturbed about the senseless shedding of Ukrainian and Jewish blood in your land?'

"The thief was silent. He thought pensively for a moment and answered the Rebbe:

> 'You are right. I am sorry. Tell me, what is this freedom fairy tale that you are telling in the midst of all this gloom and despair? What good is repeating outdated myths?'

"My Rebbe beckoned him to come to the corner of the cell where he and I were standing. My Rebbe put his arm around the thief and began to chant the story of the Exodus from Egypt again. He described in great detail the origin of the Jewish people, their sojourn in Egypt, their redemption following the ten plagues, the receiving of the Torah on Mt. Sinai.

"The thief was overwhelmed by the story. He said to my Rebbe:

> 'I never knew this. Now I think I understand a little about the Jewish people. Please tell me more.'

"My Rebbe continued telling him the story of the Jewish people through the night. The thief was absorbed with the story. As dawn was breaking, my Rebbe said:

> 'It is time to open the door for Eliyahu, the prophet of peace and redemption.'

"My Rebbe stood up and recited softly:

> 'Come in peace, prophet of peace.'

"The Ukrainian thief could not control his emotions, and he began screaming:

> 'I see him! I see him! He has really come to redeem us.'

"The Russian guard, upon hearing the screaming, came running, furious that his order had been disregarded. Without asking any questions, he shot both the Ukrainian thief and my Rebbe. I survived miraculously because my Rebbe's body fell upon me when the guard shot him.

"When I was released from that Siberian jail, I made a vow that I would repeat this story at every Passover Sayder. Eventually, my story was made known in the free world. The Jewish Agency, in cooperation with HIAS, arranged for a first degree relative in Israel to write a letter of invitation to OVIR, the Soviet Immigration Office, requesting that I be permitted to emigrate. The waiting seemed interminable, but after a very long period of waiting, I finally received my exit visa. I still honor my vow. I repeat this story in memory of my Rebbe and the Ukrainian thief each year.

It is my way of paying tribute to those who work untiringly for the redemption of their brothers from Russia, the land of twentieth century Jewish bondage."

1 This story was told to me by a friend who was a guest at the Passover Sayder in the home of the person who lived through this experience.

REPENTANCE
תְּשׁוּבָה

CHAPTER FOUR

"Sincere repentence reaches up to the very seat of G-d: the welfare of the world rests upon it."[1]

Throughout Jewish history, "teshuva" (repentance) has significantly affected Jewish life. It has helped those who may have strayed from their Jewish roots come to a reawakening of their Jewish heritage and traditional observance.

Tradition is replete with references that the concept of "teshuva" was created before the creation of the world, for G-d knew that "teshuva" was essential for the orderly functioning of the universe.[2]

"Teshuva", the regret of wrongdoing, followed by the affirmative action of seeking forgiveness by amending evil ways is one of the 613 commandments found in the Torah.

> "For you will listen to the voice of Your G-d, to keep His commandments and His statutes, which are written in the Torah, and you will repent unto G-d with all your heart and all your soul."[3]

Repentance is accomplished through fasting, tears, and the commit-tment never to return to that act of wrongdoing again. The gates of repentance are always open to man's confession of wrongdoing; for with a contrite heart, he is able to approach the Heavenly Throne, because G-d always accepts repenters.

King David taught the world the meaning of repentance when he proclaimed:

> "Wash me thoroughly from my iniquity,
> And cleanse me from my sin...
> For I know my transgressions, and my sin is ever before me,
> Create me a clean heart, O G-d,
> And renew a steadfast spirit within me,
> Cast me not away from Your Presence,
> And take not your Holy Spirit from me."[4]

The prophets urged the Jewish people individually and collectively as a nation to repent. Isaiah, the great prophet of comfort, announced:

> "Let the wicked forsake his ways, and the unrighteousness man his thought; let him return to the Lord that He may have mercy."[5]

The stories in this chapter depict how teshuva affected the lives of a number of individuals, and directed them upon the path of righteousness.

1 Talmud Bavli Yoma 86 a
2 Talmud Bavli Nedarim 39 b
3 Deuteronomy 30:10
4 Psalms 51:4,12,13
5 Isaiah 45:6

THE APOSTATE WHO REPENTED[1]

Imposition of Roman rule on the Jewish people brought about an ideological conflict between two separate cultures which was intensified after the destruction of the second Holy Temple, around the second century of the common era. Judaism and the Jewish people who believed in monotheism, ritual observance, Torah learning and the transmission of ethical values as the source of Jewish survival were threatened by the overwhelming influences of Roman paganism and Hellenism that had been absorbed into the lifestyle of the foreign masters who ruled Eretz Yisrael.

Some Jews whose faith had been shattered by suffering after the destruction of the second Holy Temple were attracted to this pagan lifestyle. It was easy for them to adapt.

Elisha ben Avuya was born during this period of struggle between Judaism and Hellenism. His father was a very wealthy man, and he invited his powerful and influential friends along with the leading Rabbis to celebrate with him the occasion of his newborn son's circumcision.

The guests sang and danced. Avuya noticed that Rabbi Yehoshua and Rabbi Eleazar were so immersed in the study of Torah while the other guests were feasting that it appeared as if they were surrounded by a "heavenly fire". When Avuya perceived the power of concentrated Torah study, he resolved to raise his son dedicated to Torah.

However, Elisha grew up in an environment permeated with the conflict between Hellenism and Judaism. He studied Torah and became a great scholar, but chose to follow the Hellenistic way of life. It was only a moment before his death that he admitted the error of his ways and repented.

When Elisha ben Avuya was a young boy, he saw a father ask his son to take a ladder, place it next to a tree, climb it and remove a mother bird from her nest before taking the birdlings to slaughter in fulfillment of the biblical injunction:

> "If a bird's nest chance to be before you in the way, in any tree or on the ground, with young ones or eggs, and the mother is sitting upon the young, or upon the eggs, you shall not take the mother with the young; you shall let the mother go, but the young you may take for yourself; that it may be well with you, and you will have length of days."[2]

The child obeyed his father; but as he descended the ladder, he slipped, fell, broke his neck, and died. Elisha was puzzled by the occurrence. He thought:

"The Torah promises a reward for the fulfillment of the commandment of obedience to parents to 'length of days'. The Torah also promises a reward for removing the mother bird from the nest before taking the birdlings to be 'length of days'. How could this tragedy have occurred contrary to the reward promised in the Torah?"

Had Elisha ben Avuya consulted his teachers, he would have been told that the reward of this biblical injunction meant life in the world to come, but he did not. Instead, he began to question the validity of his faith, and he delved more and more into the strange culture that surrounded him.

Later on, Elisha ben Avuya witnessed the massacre of the "Asarah Harugay Malchut" (the ten holy martyrs who died "al kiddush hashem" for the sanctification for G-d's Holy Name). When Rabbi Chutzpit Hameturgeman was slain, stoned and hanged, Elisha ben Avuya was aghast at the suffering of so great a Jewish sage at the hands of powerful pagans. He was bewildered by the mystery: was suffering the reward for the study of Torah?

He became an apostate. He was accepted by the Romans as an advisor in their battle to subdue the rebellious Jewish people and their religion. Elisha ben Avuya supplied them with insights that helped them formulate additional decrees restraining the observance of Judaism. He informed the Roman government on Jews who sought to perform the work they were commanded to do in such a manner that they avoided desecrating the Sabbath laws. He tried to entice young children who were studying Torah away from their studies and influence them to learn a trade: carpentry, blacksmithing, tailoring, hunting.

One day, Elisha ben Avuya approached Rabbi Mayer, who had been his former student, and asked him the following question:

"What is the meaning of this verse?
'Gold and crystal cannot equal it:
Neither the exchange of it shall be for jewels of fine gold.'"[3]

Rabbi Mayer answered:

"This means that there is nothing in this world which is equal to Torah."

Elisha said:

"My teacher, Rabbi Akiva, explained this verse to mean that just as gold and silver vessels which are damaged can be melted down and recast, so a scholar who has gone astray can repent his way through the study of Torah and good deeds."

Rabbi Mayer answered:

"If this is so, why don't you repent?"

Elisha ben Avuya exclaimed:

> "I can't, because I heard a secret heavenly voice utter that the repentance of every wayward son is accepted, except the repentance of "Acher" ("the other one"). (This was the name given to Elisha ben Avuya.)

One Sabbath afternoon, Rabbi Mayer was teaching in the study hall in Tiberias. Elisah ben Avuya rode his horse through the streets of the city. Rabbi Mayer noticed Elisha ben Avuya riding by. He emerged from the study hall and called after him:

> "You are such a great scholar, haven't you yet come to the realization that you can repent your ways?"

Elisha retorted that he didn't have the power to change his life-style.

Many years elapsed and the health of Elisha ben Avuya failed. He was bedridden.

Rabbi Mayer went to visit his former teacher to plead with him to repent his ways.

> "Now that your time to leave this world has almost arrived, won't you consider repentance?" pleaded Rabbi Mayer.

Elisha ben Avuya thoughtfully considered Rabbi Mayer's request. He said:

> "Let me clarify this. Do you mean that if I repent one moment before I die, my repentance will be accepted?"

Rabbi Mayer answered:

> "Don't you remember the verse:
> 'You turn man to contrition
> And say: Return children of men.'"[4]

Rabbi Mayer's words entered the soul of Elisha ben Avuya. He began to sob contritely. As he was sobbing, his soul departed from his body.

Rabbi Mayer concluded:

> "I am certain that my teacher repented before his death."

1 Talmud Bavli Chagiga 15 a, 15 b
2 Deuteronomy 22:6,7
3 Job 28:17
4 Psalms 90:3

THE REPENTENT TAILOR[1]

From the beginning of Jewish settlement in Poland, around the tenth century, the Jews were instrumental in aiding the economic development of the country. Slowly, the Jews became more secure in their new home, knowing that they were living in a country that welcomed them and even encouraged immigration from western Europe and the Germanic lands. At first, Rabbis and teachers were imported from western Europe to lay the foundation of religious life in Poland. By the fifteenth century, the quality of religious life had developed to such an extent that the native generation excelled in Torah commentary and Talmudic scholarship. Learning and living a uniquely Jewish life were the highest priorities in the individual Jewish communities in Poland.

The "Kahal" (the autonomous governing council of each city with a sizeable Jewish population) controlled much of Jewish life. It was responsible for the upkeep of the synagogue, the hospital and the schools. In addition, it supervised the religious instruction and the charitable institutions of the city. The "Kahal" was also an agency that distributed welfare benefits, caring for the poor, the homeless and the orphan. In general, it was responsible for the well being of every Jew.

This story is about a tailor who tried to find a solution to a personal problem. The "Kahal" of Lemberg, Poland, reacted by ostracizing the tailor whose solution did not conform to the religious standards of the community. Only through repentance was the tailor permitted to become an integral part of the Jewish community of Lemberg again.

A poor tailor and his wife lived in the city of Lemberg. The tailor was not an expert craftsman, so people did not buy the garments he sewed. He never thought of improving his skill so that more people would patronize him. Instead he grew more bitter each day about his impoverished condition and blamed all his woes upon the Jewish community.

His bitterness was balanced by his natural wit. He was an excellent story teller. The "poretz", the local landowner, heard about the tailor's stories and visited him often in his tailor shop to listen to his stories. One day, the tailor suggested to the "poretz" that he engage him as his court entertainer in his mansion on top of the hill.

The "poretz" agreed to try out the tailor as an after dinner entertainer. At first, the tailor would perform his act and return home every night. However, he gradually began to eat from the "poretz's" table, drink the "poretz's" wine, and sleep in the "poretz's" mansion. It did not take long for the tailor to become a drunkard. On the infrequent occasions when he would return home to sleep off his drunken stupor, he would invariably beat his wife and return the next day to the "poretz's" mansion. Upon learning of his nasty behavior, the Jews of Lemberg ostracized him, and the tailor became a social outcast.

One Friday afternoon, the tailor looked out the window of his little room in the "poretz's" mansion, and he noticed great excitement at the inn in the valley below. The innkeeper was directing furniture carriers and food vendors into the inn. The tailor watched in astonishment for a while, wondering about the excitement. Suddenly, he saw a carriage approaching in the distance. As the carriage drew nearer, he noticed a distinguished looking Rabbi sitting in the front with the driver while ten young men were enjoying the ride from the back of the wagon.

He ran to the "poretz" and begged him for permission to spend this Sabbath in the inn. The "poretz" was surprised by the tailor's unusual request. He said:

> "Isn't my house, my food, and my wine good enough for you?
> Why do you want to go to the inn?"

The tailor said to the "poretz":

> "You don't understand. The innkeeper is having important guests. If you give me permission to go there, I shall return with new material from my observation of the guests."

The "poretz" was getting a little weary of the tailor's jokes and he liked the tailor's idea. He wrote a note to the innkeeper ordering that he be hospitable to the tailor over the Sabbath.

The innkeeper was in a quandry. He realized that he had to obey the "poretz", but he did not want to insult his distinguished guest by seating the outcast tailor at the same table. As he pondered the situation, he arrived at a plan. He set the tables for the Sabbath meals, seating the Rabbi at one end, the students divided equally on the long sides of the table, and the outcast tailor at the other end, entirely by himself. The tailor was seated facing the Rabbi.

It was the custom of the Rabbi to deliver a discourse on the weekly Torah reading after the Friday night meal. When everyone had finished eating, the Rabbi began to speak about repentance. The students wondered at this strange occurrence because the Rabbi never deviated from his custom, but nevertheless, they listened respectfully. At the Sabbath luncheon, the students expected to hear a discourse on the Torah reading, but the Rabbi continued with the theme of repentance. The tailor's eyes did not move from the Rabbi's face. The Rabbi's words penetrated the soul of the tailor.

When the Rabbi finished speaking, he announced that he wanted to be alone for the remainder of the Sabbath. He rose from the table and walked slowly to his room. The students respected their Rabbi's request for privacy and went outdoors. The tailor, however, remained seated on his chair in the dining room until the Sabbath was over.

The minute it became dark, the tailor ran to the Rabbi's room and banged on the door. The Rabbi opened it and said in a soft voice.

"I've been expecting you."

The tailor began to cry. In a sobbing voice, he told the Rabbi who he was. The tailor said:

"I am the one who transgressed all the commandments, but I am now ready to repent my ways. Please teach me the path of repentance."

The Rabbi wanted to make sure that the tailor was sincere, so he said:

"Fast for three days."

The tailor began to cry again and said:

"I know that in order to repent, I must suffer deprivation and pain. Fasting for three days will not accomplish anything. I am ready to go to any length in order to return to the path of righteousness."

This time the Rabbi was certain that the tailor was sincere in his desire to repent, so he said:

"You must fast from one Sabbath to the next and immerse yourself in the study of Torah and doing good deeds until you will receive a sign that your repentance has been accepted."

The tailor readily accepted the Rabbi's advice. He returned to the "poretz" and asked to be discharged from his job. Since the "poretz" was tired of the tailor's jokes, he agreed. Then the tailor went to the village synagogue where he could fast and study undisturbed for the week. He waited until the people gathered for morning prayers at daybreak and joined the worshippers. After the prayers, he took a book from a shelf on the side wall of the synagogue and sat down in a corner to study. The "Shamas" (beadle) was anxious to close the doors of the synagogue so that he could attend to other business, but he realized that the tailor was immersed in learning and had no intention of moving. He became very impatient and said:

"I would like to lock the synagogue, and I can't wait for you all day."

The tailor had no alternative but to reveal the Rabbi's conditions of "teshuva" to the "Shamas". The "Shamas" was very sympathetic when he heard that the tailor had to fast and study Torah uninterruptedly for a full week. The "Shamas" suggested that the tailor study during the day in the locked basement of the synagogue.

"After the evening prayers," the "Shamas" said, "I will unlock the door and you can come up to sleep on one of the benches."

The tailor agreed to the plan.

The week passed slowly. The tailor accepted his hunger and loneliness as part of the act of repentance, hoping that upon the arrival of the Sabbath, he would receive the sign that his repentance was accepted. However, it did not happen so. The worshippers came to the synagogue that Friday night, recited their prayers and went home. Even the Shamas, who remembered to open the basement door each evening, failed to do so on this Sabbath eve. The tailor was left in the locked dark basement of the synagogue. Fearing that his repentance had not yet been accepted, the tailor began to pray with great anguish. He prayed from the depth of his soul. After an hour, he fell into a deep sleep. He dreamed that Eliyahu Hanavi appeared and told him to stop fasting because his repentance had been accepted.

> "The sign that this is so," continued the prophet, "is that within one hour the "Shamas" will come and free you from the basement room. Furthermore, after the Sabbath, sell your house in the village and buy a new one near the 'Mikvah' (the community ritual bath house). When you have done this, I will appear to you each night to teach you Torah."

The tailor woke up, and when he recalled the dream, he was relieved of his anguish. When he heard the "Shamas'" footsteps approaching within the hour, he knew the dream was true. The "Shamas" unlocked the basement door and said to the tailor:

> "I was about ready to fall asleep when I realized that you were still locked in this basement. I quickly put on my clothes and ran here because I feared you were in great danger."

The penitent tailor walked quickly to his house and rapped on the door. His wife awakened to the sound of the knocking. The Sabbath candles had already burned out and she hesitantly asked who it was, for she could not recognize the man who stood on her threshold in the darkness. She was frightened when she heard the man say that he was her husband because she feared that he was in a drunken stupor and would beat her again. She did not want to open the door. By the sound of the fear in her voice, the tailor realized how much anguish he had caused his wife.

The tailor pleaded with his wife to let him into the house. The softness in his voice and the Sabbath atmosphere made his wife wonder if this man on her doorstep was truly her husband. Finally, she decided to open the door and let him enter.

The penitent tailor entered the little house and softly greeted his wife with "Good Shabbos". He went over to the cupboard, took down the bottle of wine and the wine cup from the shelf and placed them on the table. He recited the "Kiddush" (the sanctification over the wine) and then helped himself to some food. When he finished eating, he recited the "Grace after Meals", found a blanket in another cupboard, spread it on the floor, lay down on it and went to sleep.

His wife could not sleep all night. She could not believe that the man sleeping on the floor was her husband, the tailor, but as the rays of the sun lit the little house, she recognized him.

She wondered what brought about this enormous change in his behavior, but he told her that he could not divulge his secret until after the Sabbath. At that time, the tailor revealed to his wife what had happened to him during the past week. He also told her his plan to sell their home and buy a different one near the community ritual bath. He also made her promise that she would never reveal anything that might occur in their new home. This was accomplished.

It was the custom of the Rabbi of Lemberg to sleep only until midnight, arise, walk to the community ritual bath house with his "Shamas", bathe and then spend the rest of the night studying Torah. As the two of them were walking toward the bathhouse, the Rabbi noticed a light shining from the tailor's window. He sent the "Shamas" to investigate, and the "Shamas" reported that the tailor was sewing a garment. The Rabbi of Lemberg was impressed that the tailor had become so diligent.

This scene was repeated nightly, and the Rabbi of Lemberg began to wonder why the tailor was sewing garments at night instead of during the day. One night, he decided to personally investigate the situation. When he approached the tailor's house, the light suddenly disappeared. The Rabbi knocked on the tailor's door and the tailor opened it. The Rabbi entered the house. He told the tailor that he noticed, during his nightly visits to the "mikva", the light burning in the tailor's house, and he wanted an explanation. The tailor hesitated to reveal his secret, but the Rabbi insisted, so he started his story at the beginning and finished by telling the Rabbi of Lemberg that he studied Torah with Eliyahu Hanavi every night. The Rabbi was dismayed that the penitent tailor had this great privilege that was denied him. He asked the tailor if he could listen to one lesson with Eliyahu Hanavi and the tailor promised he would ask the prophet the next night.

The tailor asked Eliyahu the next night but the prophet consented only that the tailor might repeat the lessons to the Rabbi of Lemberg the next day. He could not join their nightly study sessions.

The Rabbi of Lemberg half heartedly accepted this arrangement for a few months, but after awhile, he began doubting if indeed the lessons that the tailor was repeating to him each day were the truth. He asked the tailor to ask Eliyahu Hanavi for a sign.

The tailor responded to the Rabbi of Lemberg the next day by telling him that the sign was that there would be no deaths in Lemberg all the time the tailor lived.

The Rabbi of Lemberg accepted the sign. He told the "Shamas" to inform him of any death in Lemberg. Each time there was a funeral, the Rabbi investigated and he found that the deceased had lived outside the city limits.

One day, the "Shamas" informed the Rabbi of Lemberg that a Jew had died within the city limits. The Rabbi tore his clothes and cried bitterly, for he knew that the tailor who had studied Torah with

Eliyahu Hanavi every night had died. The Rabbi of Lemberg personally arranged for the funeral of the tailor and he delivered the eulogy. During the eulogy, the Rabbi of Lemberg told the story of the tailor's life and the greatness he had achieved through repentance. The people of Lemberg understood that the tailor had reached the level of holiness of a "Lamed-Vav Tzadik"[2] through his repentance.

This story was written for posterity in the Diary of the Jewish Community of Lemberg.

1 M.S. Kasher. Baalay Teshuva. Maarav Publishing Co. Jerusalem. 1974. This story was adapted.

2 Op. Cit. p. 50. Talmud Bavli Succa 45 b.

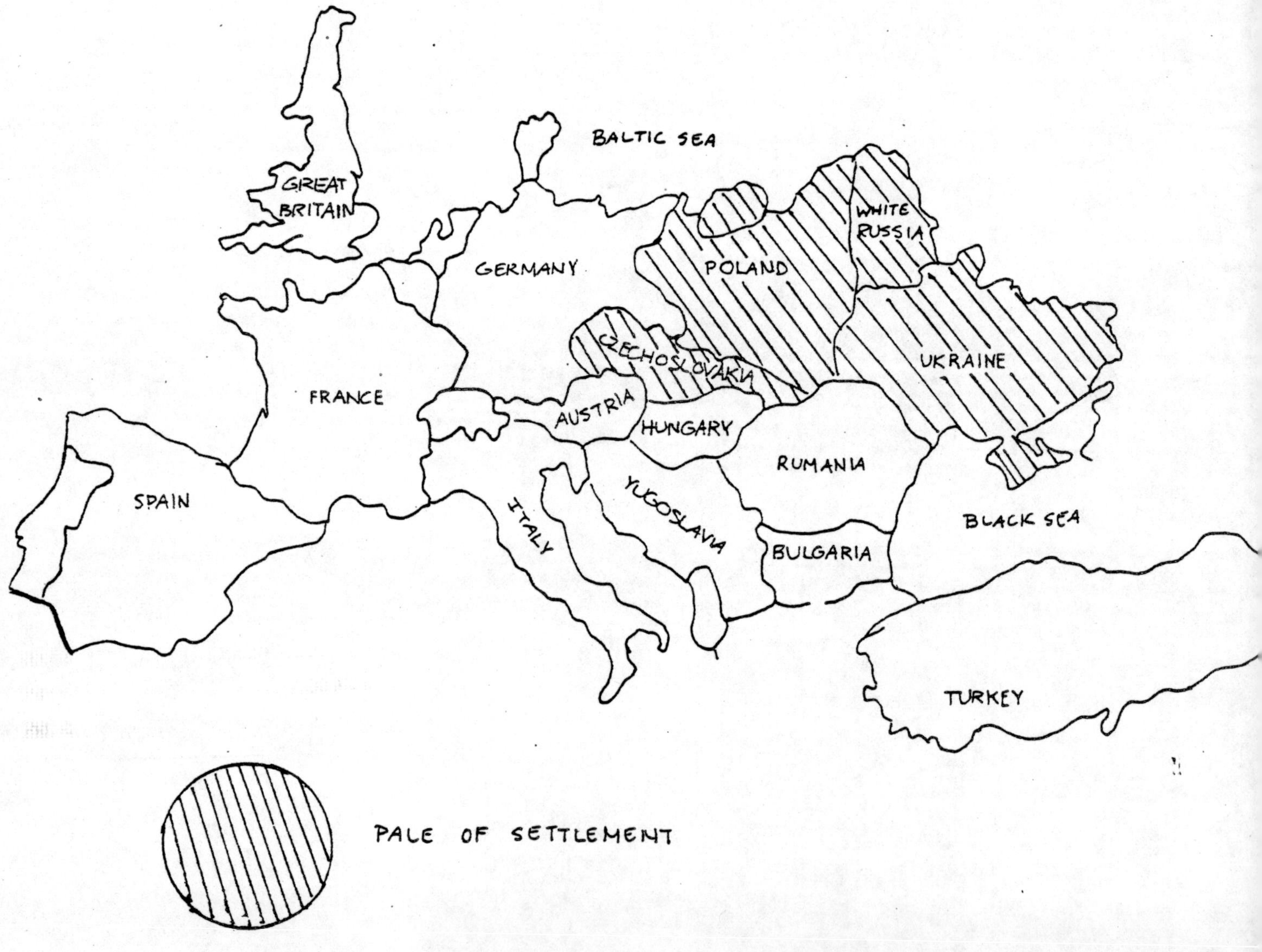
BALTIC SEA
GREAT BRITAIN
GERMANY
POLAND
WHITE RUSSIA
CZECHOSLOVAKIA
UKRAINE
FRANCE
AUSTRIA
HUNGARY
RUMANIA
SPAIN
ITALY
YUGOSLAVIA
BULGARIA
BLACK SEA
TURKEY
PALE OF SETTLEMENT

THE FORGOTTEN STORY[1]

During the eighteenth century, the Jews of eastern Europe lived in little "shtetlach", far removed from larger towns where the "Yeshivot" flourished. Since Talmudic scholarship was inaccessible, they did little learning. These Jews spent most of their time working at jobs that barely provided them a meager livelihood. Some sought to leave the poverty of the "shtetl", attempting to improve their lot in life.

For hundreds of years, Jews were accustomed to study daily. Now, this custom was replaced by weekly Sabbath afternoon lectures delivered by a "maggid" (an itinerant preacher) who travelled from "shtetl" to "shtetl". Usually the "maggid" lectured or told parables to illustrate the Torah reading for that week or discussed a pertinent religious issue.

The masses of the eastern European "shtetl" Jews were ignorant by rabbinical standards. They observed their religion out of habit rather than knowledge, with ritual rather than understanding.

It was into this spiritually impoverished environment that Rabbi Yisrael ben Eliezer, the Baal Shem Tov, the Besht, the founder of the Chassidic movement was born. He succeeded in igniting the flame of Jewish living into the souls of those Jews who might have strayed from the path of Jewish life. The Chassidic movement was instrumental in teaching the path of "teshuva", the return to G-d, in a time when Jewish ignorance was rampant.

This story tells how Shimon, one of the Baal Shem's outstanding disciples was instructed to search for a sinner and lead him on the path of repentance.

Prior to the Baal Shem's death, Shimon was given a strange assignment. The Baal Shem told Shimon to travel from "shtetl" to "shtetl", to repeat the stories, teachings, and incidents that occurred in Medziboz, the Baal Shem's hometown. The Baal Shem told Shimon that eventually he would understand the reason for this request.

Shimon heeded the Baal Shem's request. A bit sorrowfully, not knowing how long he would have to wander, he began to travel from "shtetl" to "shtetl", from village to village, from city to city.

The fame of the Baal Shem had spread throughout eastern Europe. Wherever Shimon travelled, people gathered to hear the words of the emissary of the Besht.

After a few months of travel, Shimon arrived in the town of a very wealthy Jewish noble. The townspeople told Shimon that the noble was always anxious to hear stories about the Besht. When the noble found out that Shimon was visiting the town, he invited him to visit with him for a few days, until after the Sabbath. The noble told Shimon that he would invite the whole town to be his guests for the three Sabbath meals and he could relate any incidents that he wished about the Besht.

The people of the town gathered in the house of the noble that Friday night. Long banquet tables had been set in honor of the Sabbath and the special guest. The noble recited the Kiddush; silence permeated the room. The noble wanted to hear the words of the Baal Shem from Shimon. Shimon stood up to speak but he found himself speechless. He could not recall one story, one teaching, or one incident from the Baal Shem.

Shimon sat down frustrated. The guests were disappointed but they behaved politely. They finished their meal and left the noble's house. The noble graciously comforted Shimon. He told Shimon that there were still two more Sabbath meals to which the townspeople had been invited, and he would be able to retell the words of the Baal Shem at the other meals.

The next morning, after worship in the synagogue, the townspeople gathered once again in the noble's house. After the noble recited the Kiddush, Shimon stood up to speak, but once again he was speechless. The townspeople were very disappointed. They finished their meal and left the noble's house. They did not return for the third meal of the Sabbath.

Shimon could not understand why his mind went suddenly blank. He had visited so many towns before and had no difficulty repeating the Baal Shem's teaching.

After the Sabbath, Shimon informed the noble that he had to leave. The noble, still gracious and caring about Shimon, gave him some money and a carriage to make his travels more comfortable. Shimon was very depressed, but he took the money and the carriage and set off for his next destination.

He rode towards the outskirts of the town. Suddenly, he remembered something about the Baal Shem which he thought was very important to tell the noble immediately, so he turned the carriage around and returned to the noble's house.

Shimon alighted breathlessly from the carriage and ran up the stairs. He pounded furiously at the door. The noble opened the door and Shimon said:

> "I remember a tale, and it has no ending, but I feel it is important that I tell it to you."
>
> "Come in," said the noble, leading Shimon to the study room and they both sat down by the candle-light.

Shimon spoke quietly:

> "I must tell you this story.
>
> "I was young in those years, and I travelled everywhere with the Baal Shem. The trips we made

together were incredibly distant, yet they never took much time. There were areas through which we travelled where the Jewish people were persecuted by the Church, but we were never afraid. One day we traveled to a distant town, and it seemed when we arrived that the town was deserted. The streets were vacant; the houses were boarded up; every store was closed; every window was shut; and every door was locked.

"The Baal Shem stopped his carriage in front of one large house and told me to walk up the steps and pound on the door. I alighted from the carriage, walked up the steps of the house he indicated, and began to knock, but no one answered. I returned to the carriage; the Baal Shem flung open the carriage door and he himself got out. This time the door was opened a crack and he demanded to see the owner. The woman of the house came to the door and recognized him. She pleaded with him to be quiet. She explained that this day was set aside to execute Jews in the town square, and any agitation would cause more executions.

"The Baal Shem ignored the woman's plea, pushed past her and leaped up the stairs two at a time to a third floor room. He motioned to me to follow. When we reached the top floor, he opened the window which looked down upon the town square.

"I saw over the Baal Shem's shoulder that the woman was right. In the middle of the town square was a large gallows on a platform. A shivering man sat in a corner of the platform with his hands tied behind him. Another man, dressed as an executioner, waited and looked aimlessly towards the edge of the town. A large, regal looking horseman appeared from the edge of the crowd, dressed in Bishop's clothing. He moved slowly toward the town square. As he moved through the crowds toward the platform, the eyes of the townspeople followed him.

"The Baal Shem turned toward me and instructed me to go down to the street, walk over to the Bishop, and tell him that the Baal Shem wanted to see him immediately. I did not question the Baal Shem, and trembling, I followed his instructions, walked through the spectators, and ascended the platform. I lifted my arm and pulled on the Bishop's waistcoat.

'Who are you?' the Bishop demanded.

"I pointed to the window from where the Baal Shem watched the proceedings. I said:

> 'The Baal Shem sent me to tell you that he wants to see you.'
>
> 'Later,' the Bishop exclaimed! 'Tell the Baal Shem I'm busy now, but I will see him later.'

"I ran back to the Baal Shem with the message, and in an exasperated voice, he told me to return to the Bishop and insist that he come at once. I heard the drum rolls signalling the beginning of the execution as I pushed my way through the crowd the second time. I tugged at the Bishop's coat and said firmly:

> 'The Baal Shem insists that you come immediately.'

"The Bishop ordered the execution stopped, walked off the platform, and followed me to the house and the room at the top of the stairs where the Baal Shem was waiting. The Bishop opened the door and entered. He told me to wait downstairs. That is all of the story that I know."

Shimon opened his eyes and he saw the noble swaying in his chair, tears rolling down his cheeks.

"I can finish the story. I was that Bishop.

"I converted to Christianity with the promise that I would be elevated to a high position in the church hierarchy.

"I perpetrated pogroms against my own people. Each pogrom that I planned brought with it more honor from the church. However, I was not able to sleep peacefully, because my Jewish ancestors, Abraham, Isaac and Jacob fought for the redemption of my soul. My misdeeds disturbed their eternal rest and I couldn't stand the torment any longer.

"I prayed for the Baal Shem to help me, and he began to appear to me each night as I dreamed. My sleep became more and more disturbed, so I decided to abdicate my position as bishop of the Christian Church and return to the religion of my ancestors. To camouflage my abdication, I felt I must perform the deed which you witnessed in the town square.

"It was only after the second call from the Baal Shem that I was saved from performing that dastardly act.

"The Baal Shem convinced me to repent my ways, promising me forgiveness if I would stop sermonizing against the Jewish people. He commanded me to move to this town and live alone as a businessman. He told me to wait until someone would visit me and tell me the first part of my story. This would be a sign that my sins were forgiven. Since you, Shimon, told me the first part of my story, I will divide all that is mine with you. The time has come for you to stop wandering."

1 Martin Buber. Legends of the Baal Shem. Schocken Books. New York. 1969. This story was adapted.

THE CONFESSIONAL[1]

A great change occurred in Jewish religious life after the death of the Baal Shem. The Chassidic movement which he founded spread like an uncontrolled tidal wave to the many far flung Jewish communities in the Pale of Settlement, an area comprising Poland, Lithuania, part of Russia and the Ukraine where two thirds of the world's Jewish population lived. It had a tremendous impact on the the Jewish people as it inspired them to worship with more love and to observe the commandments with more joy.

The Rebbes of the Chassidic communities created an atmosphere where both social concern for the suffering of others was shared with the study of Torah, prayer and the observance of the commandments.

One of the great charismatic eighteen century Chassidic leaders was Rebbe Layvi Yitzchak of Berditchev. He was a great and holy soul and an eloquent advocate for the Jewish people. Rebbe Layvi Yitzchak of Berditchev loved G-d and Judaism, but his love for the Jewish people was as great as his love for both combined. He refused to believe that any Jew could do wrong. He attributed the boasting of robbers to confession of sin, the admission of felons to truthfulness and contrition over transgression to a plea for compassion.

The following story, one of the most famous about Rebbe Layvi Yitzchak, occurred shortly before Rosh Hashana. The people of Berditchev had gathered in the synagogue to recite the "selichot service" (prayers of forgiveness) which is the practice at that time of the year.

It was the custom that when the people gathered and were ready to begin praying, the "shamas" (beadle) would go to inform Rebbe Layvi Yitzchak and he would hasten to the synagogue. On this particular "selichot" night, Rebbe Layvi Yitzchak accompanied the "shamas" as usual, but between the Rebbe's house and the synagogue a sudden squall unleashed furious winds and torrential rains. They searched for a place of shelter under the roof of a deserted cellar.

The "shamas" noticed a group of Jewish scoundrels sitting in the deserted cellar, eating, drinking and shouting excitedly, so he told the Rebbe. Rebbe Layvi Yitzchak rebuked the "shamas" for speaking evil about other Jews. He explained that the reason they were eating and drinking was so they might have strength to repent. Then the "shamas" noticed that the scoundrels were dividing their money. When he informed the Rebbe of this, he was again silenced. This time, Rebbe Layvi Yitzchak explained to him that they were counting their money to determine which share of their wealth they would give to charity. When the "shamas" heard them bragging about their method of operation and their great success as robbers, he reported this also. This time instead of interpreting the scoundrel's bragging, Rebbe Layvi Yitzchak ignored the inclement weather and ordered the shamas to accompany him directly to the synagogue.

When he reached the synagogue, Rebbe Layvi Yitzchak ascended the pulpit and said:

> "My holy Jews, I must tell you something before we begin the "selichot" services. The reason for my late arrival here tonight is that I was taking refuge from the storm in a deserted cellar when I encountered a group of Jewish scoundrels boasting of their success as thieves. They were confessing their wrongdoings. This made me realize that I too must make my own confession before the Almighty.
>
> Therefore, let us all join in the "selichot" services by first confessing our wrongdoings, whatever they may be, as an act of repentance."

Rebbe Layvi Yitzchak of Berditchev began to lead the congregation in the recitation of the "selichot" service. All present joined him. The people of Berditchev, realizing the saintliness of their Rebbe, understood that the moment had arrived for true repentance.

1 Yisrael Bekmeister. <u>Sipuray Niflaot Migdolay Yisroel</u>. Ayshel Press. Benay Brak. 1969. This story was adapted.

AN UNUSUAL TORAH SCHOLAR[1]

During the Age of Enlightenment, in the seventeenth and eighteenth centuries, western Europe emerged into a modern society, fostering the development of representative government, social justice and the rise of nationalism and capitalism.

In France and Germany, Jews were lured with the promise of emancipation and citizenship. They developed a taste for European culture and political rights. One group modified the prayer service, rewrote the prayer book, added choral singing and the accompaniment of the organ to the service, and advocated preaching sermons in the vernacular. They believed that Judaism and its institutions were subject to change and could be adapted to a modern society. Another group proposed that Jews must raise their spiritual standards and strive for moral perfection through the guidance of Torah, consciously fulfilling the commandments and teaching the world the Jewish prophetic mission which could only be achieved through spiritual separateness. Both groups wanted emancipation: one hoped to achieve it by conforming to society through modification of Judaism; the other by maintaining full stature as Jews.

A third group completely broke down the barriers which ghetto isolation had imposed upon the Jews for a thousand years. They began to estrange themselves from Judaism and strayed from traditional Jewish life and observance. Some intermarried in the hopes of achieving acceptance outside of Judaism.

The following story about Dr. David Bernard, a young man who temporarily left his roots in search of professional success, illustrates that the soul of every Jew is permeated with sparks of holiness that can never be erased.

A baby boy named David Bernard was born to a poor but observant Jewish family in a small town in the Pale of Settlement. The little town did not have a school and the parents could not afford to bring a teacher from another town to teach their son. They were gravely concerned that their bright child was growing up with a very limited knowledge of Jewish life. In their great anguish, David's parents decided to write to Mr. Bernard's brother, a rich and assimilated Jew who lived in Berlin. They explained to him that the town in which they lived did not have an adequate school for their son. They asked him if he would help them financially so they could send their son to a "Yeshiva" in a larger city. The uncle agreed to help educate the youngster on his own conditions, namely that the boy be sent to Berlin to live with him. He withheld his intentions to educate the lad in an exclusive secular school.

David Bernard went to live with his uncle in Berlin. His uncle enrolled him in the finest secular school and showered him with every conceivable luxury. To appease his brother, he hired a tutor to teach the boy a few prayers and rudimentary Biblical passages, but the little he learned was meaningless because there was no reinforcement in the home. It did not take long and David forgot all his father had taught him.

When David was old enough to choose a career, he decided on medicine. His uncle, influential in political circles, arranged without difficulty for David's admission into medical school. David learned eagerly and graduated with honors. He soon became a very famous doctor. As his reputation spread, he was consulted by kings and princes. He often travelled great distances to attend to the seriously ill.

David Bernard married a young woman named Margaret who had a background similar to his. One day, he received an urgent message to attend to a personal friend of Napoleon's who was seriously ill. The patient was living in a distant town. Though he started out immediately, Dr. Bernard arrived too late to save the patient's life.

He decided to return home along a road which ran through the main street of a small village. He noticed throngs of people standing outside the village synagogue. Dr. Bernard was not aware that this particular day was the Day of Atonement, the holiest day in the Jewish calendar. From inside the synagogue, he heard the voice of the Cantor chanting a soul stirring melody to the prayer: "Chay vekayam nora u-marom v'kadosh" (Our G-d is living, awe inspiring, most High and Holy).[2] The melody penetrated the depth of his soul. He was overcome with a desire to enter the synagogue. He dismounted his horse, tied him to a post and entered the synagogue. He found a place near the door and sat down.

Moments later, a man walked to the center of the pulpit and interrupted the Cantor's chanting. He pleaded for the services of a doctor from amongst the congregants to assist Rebbe David Lelever's daughter in a very difficult birth. Dr. Bernard volunteered to deliver the baby, and he saved the mother's life.

Afterwards, Rebbe David Lelever gratefully approached Dr. Bernard and said:

> "Because you returned to me the soul of my daughter and my new grandson, I will return your soul to you."

The Rebbe insisted that Dr. Bernard stay with him for the remainder of the Day of Atonement and they sat side by side on the pulpit, the Rebbe teaching, and Dr. Bernard listening. The Rebbe invited Dr. Bernard to share the "break of fast" meal with him at the conclusion of the services and to remain in his home overnight.

Early the next morning, Dr. Bernard took leave of Rebbe David Lelever. As his horse trotted slowly on the road toward his home, Dr. Bernard pondered all that had happened to him during the past twenty-four hours. He realized that the Rebbe had awakened within him his Jewish feelings. He resolved to begin searching for his Jewish identity.

As he approached his home, he noticed that Margaret was sitting on a rocking chair on the porch, deep in thought. At first he wondered why his normally busy wife was sitting idle in the middle of the day. He greeted her by saying that he had a very serious matter to discuss with her. She answered that she too had something on her mind to discuss with him.

Dr. Bernard sat down on the porch steps next to his wife's rocking chair and told her what happened to him on his trip. When he finished, she told her story:

> "Yesterday," she said, "I was very busy with the household chores, but around noon, I suddenly heard a piercing sound which seemed to penetrate the depth of my soul. I was able to decipher the sound 'Chay vekayam nora u-marom v'kadosh' and I have been repeating these words over and over since, because I did not want to forget them.
>
> "Today I realized that these words were calling me to return to my Jewish roots."

Together, Dr. David Bernard and his wife Margaret began to delve into the study of Torah. They became disciples of Rebbe Yaacov Yitzchak, Hachoze (the Seer of Lublin) with whom they studied each day. As a result of their studies, they consecrated their lives to the observance of the mitzva of hospitality to strangers and they became known for their kindness and love of other human beings.

Though David and Margaret Bernard had received only a secular education, they immersed themselves so enthusiastically in their Jewish studies that the spark of holiness which permeates the soul of every Jew was rekindled.

When Rebbe Yaacob Yitzchak, Hachoze (the Seer of Lublin) died, his followers asked Dr. Bernard to serve as their Rebbe, but he refused. He preferred to retain his medical career. Margaret so inspired Jewish women with her hospitality that the Jewish community honored her by naming their daughters for her.

1 There are many versions of this story. Two sources are:
1) M.S. Kasher. Baalay Teshuva. Maarav Publishing House. Jerusalem. 1974.
2) Niflaot Tzadikkim. Ha-admor Harofay. Published in Jerusalem. This story was adapted.

2 This verse is the refrain of a "piyut" (prayer poem), recited in the repetition of the "Amida" (the silent devotion) during "Schacharit" (morning service) on Rosh Hashana and Yom Kippur. The "piyut" begins with 'Yimloch Hashem L'olam...V'atah Kadosh Yoshayv Tehilot Yisrael...Atah hu Elohaynu Bashamayim U'Vaaretz.' (The Eternal shall reign forever...You Holy One, listen to the prayers of Israel...You Who are Master over Heaven and Earth.)

VIGNETTES OF THOSE WHO HAVE RETURNED[1]

The length of the Viet Nam war, the high number of United States casualties and the exposure of American involvement in morally unjustified crimes such as the massacre at Mai Lai helped turn many United States citizens against the war. American dismay at the atrocities of guerrilla warfare fostered draft resistance movements and huge public protest demonstrations on college campuses across the country.

Among the protestors were middle class Jewish youngsters who were imbued with the liberal philosophy of individual freedom, integration of the races, sexual equality and the eradication of poverty.

When the war ended, some of these Jewish youngsters realized that they had been raised to recognize that they were born of the Jewish faith, but they hadn't the vaguest idea of what that faith entailed. They were ignorant of Jewish practices and ritual. They had little Jewish education. They began to delve into the meaning of their Jewishness by exploring the differences between themselves and their contemporaries who were not born into the Jewish faith. They wondered about Judaism's philosophy of the sanctity of life, the meaning of Jewish existence, the problem of good and evil, the place of the Jewish people in history, Eretz Yisrael, Shabbat, Kashrut and the study of Torah. Oftentimes, when these young people turned to their families for answers to these questions, they received no satisfaction, because their families were unconcerned with Judaism as a religious tradition. These Jewish youngsters began to search out the eternal truths of their Jewish heritage, the religion into which they were born. These "Baalay Teshuva" (young people who returned to their Jewish heritage) wanted to know what "being Jewish" meant.

The following stories are a sample selected from thousands of stories about youngsters who have returned to their roots, whose "pintele yid" (the spark of their soul) is shining. I have had the privilege of knowing some of them personally. I have changed their names to protect their privacy.

Shmuel was a drop-out from society. He abhorred the pseudo-liberal values of the sixties along with his middle class Jewish upbringing which was void of religious ritual and learning. He was a young man in search of himself.

He floundered around the west coast of the United States for a while, then decided to settle down among the redwoods in the forest north of San Francisco. Strange as it seems, he carried his Bar Mitzvah "tefillin" (phylacteries) with him throughout his wanderings and considered them a cherished possession although he did not know how to use them.

One day, he decided to hitchhike into San Francisco. He had his tefillin in his backpack. The driver of the car that picked him up on the road drove through the Jewish neighborhood of San Francisco.

Something clicked inside Shmuel's head. He thanked the driver for the ride and got out of the car. He found himself standing in front of a Jewish bookstore. Shmuel entered the store.

The proprietor was a kindly man. He was not very busy, so he began to make conversation with Shmuel. They talked for a while, then Shmuel felt confident enough to reach into his backpack and remove his "tefillin". He said:

> "I have had these since my Bar Mitzvah. No one ever taught me how to use them. Won't you please show me how to use these "tefillin" and explain to me why I was given them at my Bar Mitzvah?"

The proprietor took the "tefillin" from Shmuel's hand and unwrapped them. Slowly, he placed them on Shmuel's arm and head, and helped him recite the appropriate blessings. Shmuel was overcome with emotion. At that moment, he decided that he had to find out what being Jewish meant.

In those years in San Francisco, there was a House of Love and Prayer that welcomed Jewish young people who had dropped out of society. The House of Love and Prayer provided a home, a learning environment, and an atmosphere for searching youngsters. The proprietor recommended that Shmuel stop by the House of Love and Prayer. He followed the proprietor's advice and was welcomed by the young people that were in charge. It didn't take long for Shmuel to move out of the forest and become a resident of the House of Love and Prayer. He learned slowly at first, then absorbed more and more of his heritage. Gradually, he made a strong commitment to living a completely Jewish life. He married a young lady with a similar background and together they decided to make aliya.

Today, they are living on a religious Moshav in Eretz Yisrael with their children.

Miriam grew up in the southern part of the United States in a large metropolitan city. She was a typical Jewish American teenager in that her educational background included public high school, Sunday school and afternoon Hebrew school. Her hobby was operating a ham radio station. She was very involved in her synagogue youth group and with the guidance of the youth leader, she organized an extracurricular activity at her high school, a ham radio club. She named the station WHKBH ("Hakadosh baruch Hu", blessed be the name of G-d), and it featured Israeli music and news. At least half of the students in the high school were Jewish, so the radio station had an enthusiastic following.

Among the activities of the youth group were Shabbatonim, special group celebrations of the Sabbath. Somehow, this group activity impressed Miriam a lot. She began to look forward to each Shabbaton, realizing more and more how her family's lifestyle differed from the lifestyle which the group shared at the various Shabbatonim.

The Rabbi of Miriam's synagogue was very friendly, and he encouraged the teenagers of the youth group to drop into his home on Sabbath afternoons. Miriam visited the Rabbi often.

One Sabbath afternoon, Miriam casually announced that after her high school graduation, she planned to attend the local state university. There were many young people sitting around the living room that afternoon, and a heated discussion ensued. Most of the youngsters were convinced that anyone seriously trying to find out about Jewish living should not attend the state university.

A seed was planted in Miriam's mind. She began discussing with her parents the possibility of attending a "yeshiva" for women after graduation. At first, they were opposed, but eventually they gave Miriam permission to fly to New York to see if she could find a women's learning program where she might study the depths of Judaism.

Miriam spent four years at Stern College, immersing herself in both religious and secular studies. Then she went on to graduate school. Today, she is very involved in Jewish education and outreach programs. She is trying to inspire other youngsters to become committed Jews.

Mayer was born in a suburban area of a large midwestern city. As he grew, it became easy to recognize the congeniality of the young boy who was rapidly approaching manhood by the broad "twang" of his accent and his pleasant personality.

His parents were unsophisticated and their outlook toward life in general was casual, nevertheless, they set very high standards of achievement for their son. Mayer was propelled through high school, undergraduate school and law school as if by an unknown force that pushed him from behind. Practicing law gave him no satisfaction. In fact, Mayer found the career for which he had spent so many years studying very boring.

He decided to spend a summer in Eretz Yisrael, and he discovered the excitement of a people rebuilding their homeland. One morning, Mayer walked to the "Kotel". He was awed by the thousands of people that flocked to this holy site each morning to pray, but he did not know how to join them. He moved around the praying people, trying to figure out what a Jew does at the "Kotel".

A Rabbi from one of the "yeshivot" in Jerusalem noticed Mayer's bewilderment and walked over to him. The Rabbi greeted Mayer and began speaking to him. Then, he invited Mayer to visit his "yeshiva". Something attracted Mayer to that Rabbi and that "yeshiva". Mayer accepted the invitation but the visit turned into a prolonged stay. As the summer passed, Mayer became more intensely involved in ritual and Judaic studies, things which heretofore never had meaning for him.

Soon the autumn winds began blowing in Jerusalem. Mayer had to decide whether to return to his hometown in the United States or remain in Eretz Yisrael. He decided to remain in the "yeshiva" because he was fascinated with the learning and the new observant lifestyle he had adopted.

Mayer spent a few years in the "yeshiva", but in back of his mind, he knew that he would have to support himself one day. He was certain that he did not want to practice law in Eretz Yisrael.

Interestingly, whenever he showed his class notes to friends, they complimented him on the beauty of his handwriting. They encouraged Mayer to become a "sofayr" (a scribe who writes Torah scrolls, "megillot", "tefillin", and "mezuzot"). Mayer liked the idea, so he apprenticed himself to a master "sofayr" and studied with him until he became proficient in his art.

Today, Mayer is a master "sofayr" himself. He busily practices his art, writing "megillot", "tefillin", and "mezuzot" in order to earn a livelihood. Recently, he was commissioned to write his first Torah scroll.

Mayer lives in Jerusalem with his wife and children. The only time he interrupts his holy work is when he is called to his post as a reservist in the Israeli Defense Forces.

1 This is an original story by the author.

LOVE OF MANKIND
אהבת הבריות

CHAPTER FIVE

"Love your neighbor as you love yourself, I am the L--d." [1]

"Every Jew is responsible to care for every other Jew." [2]

Although "ahavat habreyot" (love your neighbor as you love yourself)[3] is one of the 613 "mitzvot" of the Torah, it has ramifications far beyond the simple written words. Loving our fellow man implies genuine caring, devotion, and sincere involvement in the joys and happiness of our relatives, friends and neighbors as well as supporting them in their time of anguish or misfortune.

Hillel summed up the concept of "ahavat habreyot" when he stated:

> "What is hateful to you, do not to your neighbor, that is the whole Torah, while the rest is commentary; therefore go and learn it."[4]

Rabbi Akiva taught that "ahavat habreyot" is the fundamental principle of the Torah.[5] Our sages explained this principle to mean not stealing or trespassing a neighbor's property, not coveting a neighbor's wife, not inflicting personal damages on him, not embarassing him. The positive side of this principle is praising our neighbor's deeds, guarding his property, treating him compassionately, defending his honor, and answering his pleas for help.

It is incumbent upon us to act kindly because every human being needs to be treated with kindness. Examples of acts of kindness include: visiting the sick, comforting mourners, providing dowries for brides, arranging funerals and inviting guests to one's home. Our sages also included in the list of acts of kindness: treating friends with honor, inquiring after their well being, helping them in time of trouble, greeting them warmly when they visit our home, judging them meritoriously, going out of your way to do them a favor and lending them money when necessary.

Furthermore, in applying this principle insofar as it affects the Jewish people, our Rabbis interpreted this to mean not only loving our fellow Jews, but also being responsible for their needs. This means that the actions of individual Jews reflect upon the entire Jewish community, and we must be judicious in all we do.

Loving or being responsible for another human being implies understanding his needs. Rebbe David Lelever illustrated this by the following story.

"One day, Rebbe David Lelever told a student that he overheard two peasant friends discussing how much they loved one another.

> The first peasant asked: 'Do you love me?'
>
> His friend answered: 'Of course I love you, I love you very much.'
>
> The first peasant continued: 'Do you really know what I need?'

> His friend countered: 'How can I really know what you need?'
>
> The first peasant replied: 'Then you really don't love me. If you really loved me, you would know all my needs.'"[6]

G-d, in His infinite Wisdom, teaches His children how to love one another and how to live harmoniously and peacefully. This enables us to share each others' sorrows and joys and makes us aware of the continuous responsibility each Jew has to every other Jew. We are, after all, one people, who worship one G-d, Who gave us One Torah.

The stories in this chapter depict the way some people "love their neighbors as they love themselves."

1 Leviticus 19:18
2 Talmud Bavli Shevuot 39 a
3 Leviticus 19:18
4 Talmud Bavli Shabbat 31 a
5 Midrash: Sifra, Kedoshim
6 Meir Meisels. Judaism Feldheim Publishers. Jerusalem.

LET YOUR HOME BE OPEN TO THE NEEDY[1]

The spiritual development of the Jewish people was emphasized by the Tannaim, the Rabbis who composed the "Mishna" (circa 300 B.C.E. until 200 C.E.). They were concerned mostly with spiritual values because they believed that spiritual values had eternal validity.

One of the great legacies which they composed was Pirkay Avot, a compilation of ethical wills that is included in the "Mishna". It conveyed to its readers rules of conduct: how to get along with foreign governments; how to preserve the social order; how to occupy life with Torah, prayer and good deeds; how to care for the poor, the widows and the orphans; how to strive for truth, justice and peace. The "Tannaim" wanted the future leaders to be imbued with "ahavat Torah" (love for Torah) as well as "ahavat Yisrael" (love of the Jewish people) because they felt this was the essence of Jewish life.

Stories about kindness, patience, consideration and compassion were legendary during Tannaitic times. One of the "Tannaim" was Hillel Hazaken, the Elder. Together with his wife, they literally practiced "ahavat Yisrael".

The following story is about the kindness of Hillel's wife to a poor groom. Hillel Hazaken was very poor when he began his career as a student of Torah. Later in life when his abilities as a scholar were recognized, he was appointed to the presidency of the "Sanhedrin" (the Jewish Supreme Court). Although he was materially more comfortable, he still led a very simple life. In keeping with the tradition of modest living, his wife prepared the food for the family herself.

One day, Hillel invited an important guest to his home and he asked his wife to prepare a special meal in his honor. She cooked and baked making certain each dish was perfect in texture and taste. As she was preparing to serve the meal, she heard a knock at her kitchen door. She opened the door and found a poor man standing on the step. She asked the man what he wanted.

With tears in his eyes, he hesitatingly said:

> "Today is my wedding day. Both my bride and I are very poor. I have nothing to serve the guests who come to the wedding and I am ashamed to go to the wedding with empty hands. The bride and the guests have gathered and they are waiting for the wedding ceremony to begin. Could you please help me?"

Hillel's wife took all the tasty dishes and cakes that she had worked so hard to prepare and gave them to the poor groom. The groom accepted the food as a wedding gift and joyfully went to his wedding. Hillel's wife began preparing a second meal for her husband's guest. She cooked again and she baked again.

All the while, Hillel sat with his guest and discussed Torah thoughts. The hour was getting late and the guest was hungry. Hillel wondered what was taking his wife so long to prepare the meal. Finally, she was ready to serve. Only then did Hillel ask her what had taken so long. She told him the story of how she gave the poor groom the first meal that she had prepared. Hillel praised her and blessed her deeds. He said:

> "Everything you did was in honor of G-d. The Almighty rejoices over your good deed and over the love you have shown to another human being."

1 Talmud Bavli: Edyot: Masechet Derech Eretz Perek Sheshe.

THE DISPUTATION WITH THE HAPPY ENDING[1]

One of the problems which Jews faced during the Middle Ages was the organized public debates or disputations between representatives of the Christian and Jewish communities challenging the validity of Jewish religious belief. Leaders of the Jewish community were coerced into defending either Judaism or the Talmud because church leaders believed that the Talmud was anti-Christian. The Jews were permitted to interpret passages in an effort to defend against the Christian accusations. Oftentimes, church leaders used Biblical words or phrases out of context in their attempt to disprove Judaism's theological concepts and the validity of the Talmud.

No matter the outcome of the diputations, the consequences for the Jews ranged from the public burning of the books of the Talmud to expulsion.

This story describes a disputation which fortunately had a happy ending because Yaker Di, the hero, felt such a responsibility toward the Jewish community that he willingly took up the challenge to dispute the bishop of Nuremberg in an attempt to save his people from a horrible fate.

There was a certain very wealthy, childless Jew who lived in Mainz. He prayed from the depths of his soul that G-d grant him and his wife a child. One night, the prophet Eliyahu appeared to him and said:

> "It has been decided by the heavenly court to grant your wish under the following conditions: You will have a son who will save the Jewish people from much anguish, but at his birth you will become very poor. You will have a great deal of aggravation from him; however in the end, you will have great joy!"

The man begged the prophet to allow him to think over the heavenly court's conditions for one year. The prophet granted his request to return at the same time the following year.

One year later, Eliyahu returned to the wealthy Jew to inquire about his decision. The wealthy man told the prophet that he had decided to accept the conditions. Shortly thereafter, his wife became pregnant. During the year of her pregnancy, he lost all his money.

The following year, a baby was born, and the parents named their son Yaker Di which means "more than precious". From the time of his birth, it was obvious to the parents that their child was very special.

When Yaker Di was old enough to attend school, his parents enrolled him in a "yeshiva" where he learned at an amazing speed. His classmates became jealous of his achievement.

One day, while they were playing at the edge of the river that flowed near the "yeshiva", some of his classmates deliberately pushed Yaker Di into the water. He feared his classmates would attempt to drown him at their next opportunity. Rather than swim back to the shore, he decided to float down the river to the next village and take refuge in the "Yeshiva" there.

Yaker Di was wet, tired and without shelter but the "Rosh Yeshiva" (the head of the Talmudic academy) a kind and hospitable man, invited him to eat and sleep in his home and to attend classes in his "yeshiva". In exchange for this hospitality, Yaker Di volunteered to clean the study hall every night. Yaker Di looked like a beggar and no one in the new village suspected that he had the potential to become a great scholar.

It was the custom of the "Rosh Yeshiva" to finish the day's lesson with a very difficult thought question. Many of his students fasted and prayed in order to purify their thoughts so that the correct answer would be revealed to them. Each night, Yaker Di took a few moments from his cleaning chores to write out the answer on whatever pieces of parchment were available to him and he left his answer on the Rabbi's desk.

Each morning, the Rabbi was overwhelmed by the depth and brilliance of the answers to his difficult thought questions. None of the students knew who authored the answers. The Rabbi decided to find the anonymous genius, so he placed a very thin glass upon his desk one night before he left the study hall and he waited outside.

Yaker Di entered to do his nightly chores. When he completed his chores, he walked over to the desk and wrote the answer to the day's question on a scrap of parchment, not realizing that the parchment was placed on a very thin pane of glass. The pressure of his writing broke the glass and the Rabbi rushed into the study hall and discovered that the anonymous genius was Yaker Di, the beggar boy who cleaned the study hall every night.

The Rabbi invited Yaker Di to study with him privately. Yaker Di was the most profound student whom the Rabbi had ever taught. The two of them, the Rabbi and the beggar boy, scaled unknown heights in Jewish learning for many years.

Opposite the Rabbi's house lived the head of the community who had a daughter of marriageable age. One day, he asked the Rabbi to choose his best student as a suitable husband for his daughter. The Rabbi and the president decided on a plan of action. The plan was to arrange a feast to which they would invite all the students and their families. At the feast, the Rabbi would ask a question, and the student who best answered the Rabbi's question would be able to marry the daughter of the head of the community.

The night of the banquet arrived as planned, and after the meal, the Rabbi proceeded to ask his question. Only Yaker Di knew the answer. The rest of the boys became very jealous and they accused Yaker Di of being illegitimate. Since he decided not to reveal his true identity, he had no choice but to run away again.

Yaker Di fled to Nuremberg where he found the whole Jewish community fasting and praying. The bishop had ordered the Jewish community to send representatives to a disputation, and if he were not satisfied with the answers, he would forcibly convert every Jew in Nuremberg to Christianity. Yaker Di volunteered to meet the challenge and the Jewish leaders of Nuremberg gratefully accepted his offer.

The Jewish community was in a dejected mood for they knew from previous experience that neither the questions nor the answers mattered to the bishop. However, Yaker Di's arguments were so brilliant that the bishop accepted his answers and rescinded the evil decree.

Shortly after the disputation, Yaker Di moved to the province of Bavaria. He was not only erudite in the study of Torah, but he expressed a deep interest in medicine, so he decided to enter medical school. The study of medicine was as easy for Yaker Di to grasp as the study of Torah, and it did not take long for him to become a famous doctor. His reputation spread throughout Bavaria.

Two interesting incidents occurred around the same time. The King of Bavaria issued a decree expelling all his Jewish subjects and the Crown Prince was afflicted with a nearly fatal illness.

Yaker Di heard of the Prince's illness and he volunteered to try to cure him. Miraculously, Yaker Di was able to save the life of the Prince. To reward Yaker Di, the King rescinded the evil decree of expulsion. He also arranged a big banquet in honor of the recovery of his son. He invited all the community nobles and leaders of all the towns in his province. Among the guests were the Rosh Yeshiva, Yaker Di's former teacher, and the head of the community, the father of the girl that he was supposed to marry.

Yaker Di recognized his former Rabbi and the head of the community. He inquired about his daughter and was told that she had never married. The head of the community did not remember Yaker Di. Imagine his surprise when Yaker Di finally revealed himself! The Rabbi and the head of the community decided to arrange a wedding for the couple, and the King ordered a luxurious palace to be built for them.

Yaker Di, observing the "mitzva of ahavat habreyot" offered his services freely to every poor person in the province of Bavaria.

One day, Yaker Di was called to attend a very sick old man. As he entered the old man's room, he recognized his father whom he had not seen in so many years of wandering. The father, though aged from worry and concern, was able to recognize his son and they were joyfully reunited. Yaker Di's father tearfully realized the fulfillment of the prophet Eliyahu's promise and he joyfully led his son to the bridal canopy.

1 Op. Cit. p. 37. Sippurim Prager Sammlung: Judisher Legenden.

GUARD YOUR TONGUE FROM SPEAKING EVIL[1]

By the beginning of the eighth century, the Arab Empire stretched from the Babylonian lands east of Eretz Yisrael through North Africa, Southern Italy and Spain. The vast majority of the people who lived in this sprawling empire followed the Islamic religion.

Moslems generally tolerated religious differences throughout their empire. They understood the economic advantage of inviting Jews to live in their midst, so they extended hospitality to them. The Jews gratefully accepted this hospitality because they were searching for a friendly place to settle.

The Jews became merchants and expert craftsmen in metal, leather, silk and wood. They shared with their Moslem neighbors the bountiful harvests gleaned from economic prosperity and intellectual stimulation. The Jews who lived under Moslem rule distinguished themselves as great Torah scholars, biblical commentators and poets. They were also prominent statesmen, physicians and financial advisors; and they held important positions in the government of the Caliphs, the Sultans and the Grand Viziers. This period of splendor in Jewish life was the "Tur Hazahav" (the Golden Age of Spain).

One of the prominent advisors to the Caliph of Granada was the eleventh century scholar and author Rabbi Shmuel ben Yosef. The Jewish community honored him with the title of "HaNagid" (the prince). He was famous for his Hebrew poetry and his Mavo HaTalmud, an introduction to the Talmud which is included at the end of most editions of Tractate Berachot.

He was the acknowledged leader of the Jewish community in Granada, Spain, serving as chief Rabbi, supporting the poor and scholars alike. He was also highly respected as minister of state in the court of the Caliph Habus of Granada. Both the Jewish religious and the Moslem secular worlds revered him. This respect put him in the position to exemplify "ahavat habreyot".

The story begins in the little spice shop which Shmuel opened as a young man in order to earn a livelihood. When he was not engaged in business, he studied Talmud or wrote poetry. Sometimes, he wrote reports for the servant of the Vizier. The reports came to the attention of the Vizier who was duly impressed with Shmuel's writing style and his calligraphy. Shmuel was invited to become the personal secretary of the Vizier. The Vizier also promoted Shmuel to being his personal counselor and followed his advice before any undertaking.

Before the Vizier died, he informed the Caliph of Shmuel's ability. The Caliph invited Shmuel to be his minister of state.

One day, the Caliph and Shmuel were riding in a procession. A large crowd lined both sides of the roadway. From the crowd, an angry Arab merchant began insulting Shmuel.

The Caliph was incensed with the man's expression of hatred and jealousy which was directed at his revered and kindly minister. He ordered Shmuel to punish the offender by cutting out his jealous tongue, so that he would never again be able to utter such angry insults at anyone.

The Caliph assumed that the punishment he decreed had been carried out. Much to his surprise, he later found out that his order had been disregarded. The occasion was a chance meeting between the Caliph, Shmuel and the Arab merchant. The Caliph was shocked when the Arab spoke, albeit respectfully. The Caliph turned to Shmuel and asked him why his decree had been disregarded. Shmuel said:

> "Your eminence, I did not disregard your decree. I did cut out the merchant's jealous tongue. I also replaced it with a gracious tongue."

Rabbi Shmuel Hanagid demonstrated how to "love your neighbor as you love yourself".

1 Nathaniel Kravitz. <u>Three Thousand Years of Hebrew Literature.</u> The Swallow Press. Chicago. 1972.

THE GOLEM OF PRAGUE[1]

During the latter part of the Middle Ages, beginning in the mid-twelfth century in Norwich, England, the Jews of Europe faced the problem of dealing with the perpetrators of Blood Libels.

The Blood Libel was the common belief that Jews killed Christian children and used their blood to bake matzos to use for the Passover holiday (of course this libel neglected to consider the Jewish laws prohibiting murder and the consumption of blood). If a Christian child were found dead in a field around the Passover season, a pogrom against the entire community usually followed. The majority of the peasants, exhorted by their leaders to believe this false accusation, used this opportunity to wreak vengeance upon the Jews. There were never witnesses to stand in defense of the Jews. The usual result of the Blood Libel was the expulsion of the Jews from their homes or the massacre of the entire Jewish community.

One of the most infamous Blood Libels took place in the Prague ghetto. This story involves the creation of a "Golem" (a clay-shaped human form) that reacted to the command of Rabbi Yehuda Lev ben Bezalel, the Maharal, when he inserted a piece of parchment inscribed with the Holy Letters of the Eternal's Name into its mouth. Because the Maharal cared for the Jewish people like a shepherd cares for his sheep, he prepared the "Golem" as a defense should his intuition of a pending Blood Libel come to fruition. The remains of the "Golem" were hidden in the attic of the "alte-neu shul" where an aura of mysterious holiness still envelopes the place.

The fate of the Jewish community of Prague delicately balanced between King Rudolph of Bohemia and Count Batislav von Lehn near whose estate the ghetto was located.

The Jews were taxed more than double that of the rest of the King's subjects, and it was their money which supported the King's wars and his commercial enterprises.

One year, Count von Lehn borrowed a sizable amount of money from the Jewish community to support a new business venture. When his business failed and he could not repay the money that he borrowed, the Jews of the ghetto suspected that he would disclaim the loan and look for a pretext to harrass them besides. Their suspicion proved correct.

As the holiday of Passover was approaching, the Jewish community feared a Blood Libel which often occurred during this time of the year. The Maharal made plans to prepare a defense against the pogrom which he felt would soon be perpetrated against the Jews of Prague. He asked four trusted community leaders to join him that night at the banks of the Ultava River. From the wet soil of the river banks, they dug fresh clay chunks, placed them on a board and secretly carried them to the attic of the "alt-neu shul".

On "Shabbat Hagadol" (the Great Sabbath which precedes Passover) the Jewish community gathered for prayer in the "alt-neu shul". Halfway through the service, there was a great commotion at the door of the synagogue. Count Batislav von Lehn was standing in the doorway, surrounded by a frenzied mob. Disrupting the prayers, they marched toward the center of the synagogue. Count Batislav von Lehn raised his hand for silence and then he began to speak. He said:

> "I cannot find my daughter. I know that you have kidnapped her. I demand that you return her immediately."

The Maharal shuddered for he realized that this was the beginning of the anticipated Blood Libel. He knew that the Jews knew nothing of the whereabouts of the Count's daughter. The Count continued:

> "I will give you three days, until the holiday of Passover, to return my daughter unharmed."

Then he marched out of the "alt-neu shul", the frenzied mob followng him. A fearful cry spread through the synagogue and although the Maharal tried to quiet his congregation, they cried helplessly.

After the Sabbath, the Maharal took his four trusted friends to the attic where he had hidden the clay. There he shaped the chunks of clay into the form of a man and in its mouth the Maharal placed a piece of paper inscribed with the four Holy Letters of the Eternal Name. The clay shaped man or "Golem" stood up and waited for a command. The Maharal commanded the "Golem" to find the Count's daughter and the "Golem" started exploring the ghetto, searching every house, every street, and every alleyway, but the girl was nowhere to be found. After searching the ghetto unsuccessfully, it took off for the open road followed by the Maharal and his four trusted friends. Heading toward the Count's palace, the "Golem" entered the great halls and descended to the basement. The group of people who were following the Golem found an astonishing and terrible sight. The Count's daughter was laying on the basement floor, slightly clothed, hardly able to speak. At her feet lay a slaughtered pig, its blood flowing all over the damp floor. Nearby, there were empty jars with notes instructing that the jars be filled with this blood and placed in the "alt-neu shul". The perpetrators of this Blood Libel would insist that the Jews murdered the girl in order to use her blood to bake "matzos" for Passover.

The Maharal confronted the Count with the evidence of the vicious plot. The Count pleaded innocent, blaming his servants for it and ordered the mob to retreat from the gates of the ghetto and not molest its inhabitants. The Maharal saved the Prague Jewish community and the holiday of Passover was joyously celebrated.

[1] Op. Cit. p. 37. Sippurim Prager Sammlung: Judisher Legenden.

YOSSELE THE HOLY MISER[1]

Jews living in Poland during the fifteenth and sixteenth centuries were economically and politically self sufficient. There were bakers, bookbinders, weavers, craftsmen, artisants and scholars. The "Kahal" (the autonomous governing council) concerned itself with social welfare and education.

The Jews lived in such close proximity to each other that they formed a cohesive community. The "Kahal" was successful in wielding tremendous power because no one wanted to be labeled a non-conformist. Generally, the "Kahal" was able to impose its will on the rich to give enough charity to support the poor. Yossele was an exception. He did not follow the rules of the "Kahal" and therefore, he was treated like an outcast.

Yossele was the wealthiest Jew in Cracow, but people called him "Yossele the Miser". He never gave a zloty (name of a small denomination Polish coin) to charity. The people of the community would beseech him for assistance, but he did no more than ask them for their names and addresses.

Yossele grew older and the Jews of Cracow figured that his time to leave this world would arrive soon, so they appointed a representative of the "Chevra Kaddisha" (the burial society) to visit Yossele and inform him that if he did not pay his share for support of the poor of the community, the "Chevra Kaddisha" would not bury him. Yossele listened to the representative, and when he finished his plea, Yossele ordered him out of his house. The representative had no effect on Yossele's behavior. He refused to pay his obligations to the community charity fund.

Yossele died the following Monday. Two days went by and the "Chevra Kadisha" refused to make arrangements for his burial in the Jewish cemetery. Finally, a compassionate neighbor dug a grave outside the fence of the Jewish cemetery and buried the corpse.

On the Thursday morning after Yossele's death, there was a tremendous commotion in front of the Rabbi's house. A long line of people were complaining that they had no money to buy food for Shabbat. The Rabbi tried to calm them. He asked the people what they used for money the previous week, since they had never asked for assistance before. They answered in one voice:

> "We never needed money before. We always looked under our doormats every Thursday morning and found enough money to buy food for Shabbat. This week none of us found any money."

The Rabbi realized that all these people who were clamoring for money had been supported by Yossele, the "most selfish Jew" in Cracow.

With great pain in his heart, the Rabbi proclaimed a day of public fasting and prayer so that the Jewish community of Cracow might atone for mistreating Yossele, the generous, anonymous philanthropist.

After the public prayers concluded, the people left the synagogue but the Rabbi remained. He sat very despondently in his seat at the front of the synagogue, deep in thought. After a while, he fell asleep.

The Rabbi dreamed that Yossele appeared before him. The Rabbi began to plead with Yossele:

> "Please forgive me for the terrible injustice I permitted to be perpetrated against you. The community despised you for your lack of generosity and I did not insist that your body be buried with proper dignity."

Yossele answered:

> "What kind of injustice did you perpetrate against me? I was able to give charity in its highest form. I never wanted the poor of Cracow to know where their money was coming from. As far as burying me properly... there was such a tumult in heaven over who would have the privilege of burying me... Abraham, Isaac and Jacob fought for the privilege... finally, Eliyahu Hanavi, the prophet, performed the task of burying my body."

Yossele disappeared from the Rabbi's dream, and he awoke, startled.

The next morning, the Rabbi commissioned a stonemason to place a monument on Yossele's grave. He asked the mason to inscribe the monument with the words:

> "Here he lies Yossele the Holy Miser. I am a witness."

Visitors to Cracow have reported finding this monument outside the fence of the Jewish cemetery. It is a memorial to a man who truly practiced the mitzva of "ahavat yisrael".

1 According to historical tradition, the Rabbi in this story was Rabbi Yom Tov Lipman Heller, the Tosofot Yom Tov, the great commentator on the Mishna. This fact was revealed to me by Rabbi Yaacov Katz, a Gerer Chasid, the "schochet" (ritual slaughterer) in Munich, West Germany, as I sat in his living room one evening in July 1982, sharing stories with him and his wife. He informed me that he heard this story from his father, who heard it from his father, who heard it from his father.

A DOWRY FOR THE BRIDE[1]

The Chassidic movement brought charismatic leaders to the forefront of Jewish life. In little "shtetls" all over eastern Europe, Chassidic courts flourished, guided by "tzaddikim" (righteous leaders) who were imbued with inspirational emotion and religious fervor. They guided their followers to serve G-d with joy. The followers of these Chassidic Rebbes discussed with awe the deeds of their leaders. The deeds became stories, and the stories were repeated over and over until a rich folklore evolved.

Rebbe Pinchas Shapira of Koretz was a contemporary of the Baal Shem Tov, but he lived longer than the founder of the Chassidic movement. He was acknowledged as a teacher of humility and truth. In this story, Rebbe Pinchas teaches a friend how to love an undeserving Jew.

Normally, Rebbe Pinchas of Koretz's "shamas" (beadle) was an organized and responsible man, but this one morning, he was disturbed and distraught. He ran around the town perplexed and disorganized. When Rebbe Pinchas asked his "shamas" what was upsetting him, he answered:

> "The disgusting tailor, the social outcast, the one that did not go to the synagogue to pray, the one that did not patronize the kosher butcher, the one that did not do a charitable act all his life, died and no one wants to attend his funeral."

Astonished at the news, Rebbe Pinchas said in a sad voice:

> "I loved that tailor so very much, I will attend his funeral."

The puzzled "shamas" seeing Rebbe Pinchas so saddened, thought the tailor must have been one of the hidden "lamed-vav tzadikkim".[2] Thinking that he made a terrible mistake, the "shamas" ran through the town again, this time informing the people that the tailor was one of the hidden "lamed-vav tzadikkim". The people stopped what they were doing and joined the funeral procession. Rebbe Pinchas walked behind the casket, leading the procession all the way to the cemetery on the outskirts of the town.

After the funeral, a friend of the Rebbe's reproached him.

> "Rebbe", he said, "you can fool the townspeople but you can't fool me. I know that tailor was a social outcast and not one of the 'lamed-vav tzadikkim'."

Rebbe Pinchas turned to his friend and said quietly:

> "There was something special about this tailor that was unknown to the inhabitants of this community. I would like to reveal it to you now. Do you remember Rachel, the orphan girl who grew up in my house? Four weeks ago, I married her off to an orphan boy. It was my privilege to establish a home for these two orphaned children. I had enough money to buy a few necessities to furnish

their house and to pay for the marriage feast. I needed twelve rubles to buy a new 'talit' for the groom as a gift from the bride, and I did not know where I could get the money. As I walked around the streets of the town searching for a solution, I passed the tailor's house. Instinctively, I knocked on the door and as he opened it, he was overtaken with joy by my presence because no Jew in this town would have anything to do with him. As a token of appreciation for my visit, he offered me one ruble. I thanked him, but told him I needed twelve rubles to pay for the 'talit'. He replied that he had only one ruble. I took it and left. As I was walking up the street, I heard the tailor shouting after me:

> 'Rebbe, if I give you the twelve rubles you need, will you promise me a share in the world to come?'

"I promised the tailor that he would have a share in the world to come as a reward for the 'mitzva' of helping to provide a dowry for a bride. The tailor ran back to his shop, picked up the additional rubles and gave them to me.

"Now that you know the whole story, Rebbe Pinchas continued, you understand why I walked behind the casket and accompanied the procession to the cemetery. I also want you to know that as I was walking to the cemetery, I envisioned the tailor's soul being carried to heaven, wrapped in the 'talit' that I purchased with the money he gave me."

1 Sh'moneh Seforim: Tzeror Hachayim. Printing House of Rabbi Natan Neta Kronenberg. Bulgaria. 1903. Reprinted in New York, 1983. This book is an anthology of Torah discourses and stories. There is no single author.

2 Op. Cit. p. 50. Talmud Bavli Succa 45 b.

TWO FRIENDS[1]

Rebbe Menachem Mendel of Rimanov and Rebbe Naftali of Ropshitz were of the second generation of Chassidic leaders. This story, told in their names, explains the abstract concept of "loving your neighbor as yourself"[2] in a most poignant way.

Rebbe Naftali, the Ropshitzer Rebbe, was asked how he learned to love the whole world. He replied:

> "When I was sixteen years old, I was privileged to 'daven' (pray) in the synagogue of Rebbe Menachem Mendel of Rimanov on Rosh Hashana. Let me tell you what happened before Rebbe Mendel sounded the 'shofar'."

He ascended the pulpit and began speaking about the importance of 'ahavat Yisrael'. He quoted the verse from the Torah:

> "Love your neighbor as yourself, I am the Lord."

Then he asked rhetorically:

> "What does I am the Lord have to do with love your neighbor?"

To gain the interest of the congregation, the Riminover Rebbe said:

> "I understand when G-d gives us a commandment to sit in the Succah or to eat matza that the verse ends with I am the Lord. I do not understand when the verse reads love your neighbor as yourself, I am the Lord. What does love your neighbor have to do with I am the Lord?"

As if to answer his own question, the Riminover Rebbe began to tell this story.

> "Many years ago, two Jews lived in Tzarist Russia. One was named Yaacov and one was named Moshe. As children, they learned and played together. They became the closest of friends.
>
> "When they reached adulthood, they swore to each other that nothing would stand in the way of their friendship. They decided that every Friday afternoon, a few hours before the Sabbath, they would stop whatever they were doing to visit with one another. The bond of friendship between Yaacov and Moshe grew stronger, even after they both married and began to raise families.
>
> "It happened that Yaacov was not earning a decent livelihood. After much soul searching, he decided to move to another town that offered the potential of earning more money to support his family.
>
> "Yaacov was filled with personal anguish at the thought of separating from Moshe, but he knew that

he had to move to a place that had the potential for economic prosperity.

"With deep emotion, he revealed his decision to Moshe. Moshe understood Yaacov's decision. The two friends decided that they would write long letters to each other each Friday afternoon before the Sabbath as a substitute for their weekly visits.

"Yaacov and Moshe wrote letters to each other for many years. Yaacov prospered. He became a very wealthy businessman. Moshe, on the other hand, found it more and more difficult to earn a livelihood in his village. He became very poor. Moshe was reluctant to let Yaacov know of his terrible financial plight, but his wife prevailed upon him to travel to Yaacov's city and beseech his friend for help.

"It was very difficult to travel in Tzarist Russia in those days. It was more difficult for Moshe since he had no money to pay the coach fare on the train. Moshe set out for Yaacov's city on foot. He travelled many months. When he arrived there, he was totally exhausted. He found a secluded spot under a tree near the center of town, sat down, and immediately fell asleep.

"That night, someone set fire to the church. In the morning, the police began searching for the arsonist. When they found Moshe sleeping under the tree, they arrested him and accused him of the crime. It was an easy thing to do since he was both a stranger and a Jew. The arrest caused a commotion, and ten witnesses stepped forth to assure the police that they actually saw Moshe set fire to the church. The police chief ordered Moshe to be placed in a stockade so that all who passed might gaze at the culprit, spit at and kick him.

"Yaacov, who had become the wealthiest Jew in the city, heard that a strange Jew was being held in the stockade in the center of town. He wanted to see the face of the Jew who had burnt down the church and brought such disgrace and shame on the Jewish community. Yaacov even feared a pogrom would occur that would annihilate the entire Jewish population of the city. When he looked at the man in the stockade, he could not believe his eyes. He had not seen his friend Moshe for many years, but he recognized the impoverished, frightened man standing in the stockade. He could not believe his friend Moshe could do such a thing.

"Yaacov thought that Moshe's impoverished condition had affected his sanity, and he was not responsible for his actions, so he went to the police and confessed to the heinous crime. The police did not care who was guilty as long as they had someone in the stockade. They freed Moshe and put Yaacov in his place.

"Moshe wanted to leave the town as quickly as possible but he was curious to see who had taken his place in the stockade. He was appalled to recognize his friend Yaacov in the stockade. He thought that Yaacov had become insane. He hoped there might be some logical explanation for Yaacov's behavior, so Moshe went to the magistrate and confessed to the crime.

"The magistrate was puzzled. He knew there was only one crime, yet the two culprits confessed their guilt. He decided to turn the matter over to a higher court. Yaacov and Moshe were transferred from one judicial district to another, but they never saw each other face to face. All the judges who reviewed the case refused to decide in favor of either defendant. Finally, the case came before the Tzar.

"Just as they were being brought before the Tzar, Yaacov and Moshe met for the first time. Yaacov asked Moshe why he committed such a dastardly crime. Moshe asked Yaacov the same question. They both realized they were innocent. They both realized they were willing to die for each other.

"When the case was presented to the Tzar, he was filled with remorse. The Tzar said:

> 'I am the Tzar of all Russia. I have more land, more wealth and more power than anyone in the world; yet I do not have a friend like Yaacov or Moshe.'

"The Tzar turned to the two friends and said:

> 'Please, please won't you include me in your friendship?'"

Rebbe Mendel of Rimanov concluded the story by saying:

> "G-d created the whole world. To Him belongs the earth and the fullness thereof. G-d only wants to be included in friendships like the friendship between Yaacov and Moshe. Therefore, when G-d commands us to 'love your neighbor as yourself, I am the Lord, G-d is only beseeching us to include His Presence in our friendship."

A thoughtful stillness prevailed in Rebbe Mendel's synagogue as he sounded the shofar.

The Ropshitzer said:

> "On that Rosh Hashana I learned the meaning of 'love your neighbor as yourself, I am the Lord.'"

1 Eliyahu Ki Tov. Sayfer Haparsheyot: Kedoshim. Alef Publications (Ben Arza). Jerusalem. 1970. This story is adapted from a legend which explains the meaning of "love your neighbor as you love yourself".

2 Leviticus 19:18

AND YOU SHALL LOVE[1]

About three generations after the birth of the Chassidic movement in eastern Europe, "the Haskalah" (The Jewish Enlightenment Movement) began to emerge in western Europe. The French and German nineteenth century Jews were imbued with the idea of equal rights and equal citizenship. Some of them were deluded into believing that if they would be less observant and more secular, they would be accepted more easily into the civil society.

The "Haskalah" stressed the cultural revival of Hebrew as a literary language, and the study of Jewish history and religion (Juedische Wissenschaft) through critical analysis. Gradually, the study of Talmud, which had played such a major role in the development of Jewish scholarship, became secondary to mathematics, science and geography. The Jews who followed the "Haskalah" wondered if Talmudic law was still applicable, if they had to adhere to rabbinic Judaism, if it were wise for Jews to remain isolated and if the Messiah would ever arrive.

Meanwhile, the Jews, living in the Pale of Settlement under the domination of Tzarist Russia, were economically impoverished and politically restricted. They constantly feared their children being forcibly kidnapped to serve a compulsory twenty-five year term in the Tzar's army. They were compelled to enroll their children in government directed secular schools. Only a very small number were successful in escaping the harsh government ordinances and provided a Yeshiva education for their children.

Jewish life was threatened by the harsh education ordinances of the Tzarist regime in eastern Europe. Great teachers lived at this time, who devoted their lives to inspire, to raise Jewish standards of living, to influence, to mold, to spread hope, often at great sacrifice to their families and their own personal health.

Among these Torah giants was Rabbi Yisrael Salanter. He brought blessings to those who fell under the sway of his influence.

Yisrael was born at the beginning of the nineteenth century in Zhagory, Lithuania. He learned Torah and the observance of the mitzvot from his father, and kindness, respect and consideration from his mother. He was recognized as an "illuy" (a genius) at a very young age. By the time he was ten years old, he could deliver a Talmudic discourse. He continued his studies in the town of Salanty.

Rabbi Yisrael Salanter is considered to be the founder of the "Mussar" movement. The "Mussar" movement stressed the perfection of ethical conduct, the refinement of character and the awareness of one's responsibilities both as a human being and as a Jew. People involved in the "Mussar" movement strove to become G-d fearing so that they might fulfill the "mitzvot" properly. Rabbi Yisrael Salanter urged people to deal honestly and kindly with each other. He did not want to limit his influence to Lithuania so he became a travelling Rabbi. He travelled through western Europe, Haskalah territory, as well as through Tzarist Russia. He felt that the Jewish world needed to have a broader perspective of "ahavat Yisrael" (love and kindness between one Jew and another). The three parables that follow describe the lessons in ethical conduct which Rabbi Yisrael Salanter used in order to influence the actions of those people he met.

"Imagine nine men in a synagogue who have gathered to pray. They await the tenth man, and when they see him coming down the street, they knock at the windows and motion for him to hurry into the synagogue. They need him to complete the 'minyan' in order that they might begin their prayers. However, imagine nine men sitting down to a meal. Do they stand at the windows, await the passers-by and invite them to share the meal? 'Ahavat Yisrael' means that you are as concerned about the strangers who are hungry as you are with the tenth man who completes your 'minyan'. When you invite the stranger into your house to share your meal, then you have learned the meaning of 'ahavat Yisrael'."

1 Chaim Yitzchak Lipkin. Taryag Mitzvot B'Halacha U'V'Agada. Netzach Publishers. Benay Brak. 1962. This story was adapted.

A LITTLE WATER[1]

Rabbi Yisrael Salanter was once invited to dine in the house of his friend Rabbi Yaakov Karpas. There were many guests and before dinner was served, each guest washed his hands and recited the blessing. Rabbi Yisrael noticed that each guest used a lot of water while washing his hands. He used only a little.

When all the guests were seated, the host asked Rabbi Yisrael why he had used so little water.

Rabbi Yisrael answered:

"I used a mimimum of water because your house stands on a hill."

The host was puzzled at Rabbi Yisrael's answer, and said:

"I do not understand what the location of my house has to do with the minimum of water you used."

Rabbi Yisrael answered:

"Think of the water carrier."

The host still did not understand. Rabbi Yisrael began to explain.

"I noticed that your maid has to carry the buckets of water from the well below the hill. Carrying water is a hard chore. If I use less water, your maid will not have to carry so much water uphill so frequently. Washing my hands before eating is a law, but I have no right to use more water than necessary. I was trying to alleviate another human being's burden."

1 Rabbi Zalman F. Ury. The Story of Rabbi Yisrael Salanter. Torah Umesorah Publications. New York. 1971.

A WIDOW AND BAKING MATZA[1]

This story about Rabbi Yisrael Salanter showed his great concern for the welfare of a poor widow. It is a true expression of "ahavat Yisrael"

Before the spring wheat harvest, Rabbi Yisrael Salanter went to the fields and personally supervised the cutting of that portion of the wheat which was set aside for baking matzot for Passover. He would go to the bakery and watch the baker grind the wheat and prepare the dough. He supervised the entire process each year for the entire Jewish community.

One year, Rabbi Yisrael Salanter was recuperating from an illness at the time that the dough had to be prepared. His students volunteered to supervise the procedure in his place. They promised to be extremely cautious with every particular detail in the preparation of the dough for the baking of the matzot. When they came to Rabbi Yisrael Salanter for last minute instructions, he said to them:

> "There is an old widow who comes each year to bake her own matzot. Be very careful to do everything to help her and be certain you do not raise your voices to her or embarrass her in any way. Caring for the widow is as important as preparing the dough for the 'mitzva' of baking matza."

1 Ibid. The Story of Rabbi Yisrael Salanter.

THE SHADCHAN, THE MATCHMAKER[1]

Since the long bitter night of the exile began with the destruction of the second Holy Temple in 70 C.E., the Jews have yearned and prayed for redemption, for G-d to return His Presence to Zion. Throughout that time, Eretz Yisrael was ruled by a succession of various foreign powers including the Romans, the Byzantines, the Seljuk Turks and the Marmelukes. From the fourteenth to the twentieth centuries, the Ottoman-Turks ruled Eretz Yisrael and it was isolated from western European influence. During the years of World War I, the Turkish rulers lost interest in the welfare of both the Jewish and Arab inhabitants of the land. They were more concerned with the danger posed by the British presence in Egypt.

The Jewish population in Eretz Yisrael was small. The Jews were scattered among the cities of Jerusalem, Tiberias, Safed and along the coastal plain. They eked out a living as carpenters, builders or porters. A group of Jews, residing in Jerusalem, considered it their sacred responsibility to study Torah day and night in order to maintain the privilege of living in Eretz Yisrael. They were called "Shomray Hachomot" (the Watchers of the Holy Walls) and they were sustained by the charity of their brethren who resided in exile.

European Jews had begun returning to Eretz Yisrael towards the end of the nineteenth century in great numbers. They were impelled by the Zionist movement whose goal was to reestablish an independent Jewish state in Eretz Yisrael.

The early settlers who made "aliya" (going to live in Eretz Yisrael) planted the seeds which made the deserts bloom, revived the Hebrew language, built villages and towns, developed an economic system and laid the foundation of a Jewish defense force. The pioneers, together with the "yishuv" (those who were already living in Eretz Yisrael), joined forces to rebuild their ancient ancestral homeland.

Life was very difficult for both the natives and the pioneers. They had to cope with typhus epidemics, plagues of locusts and malaria transmitted by infected mosquitoes inhabiting the swampland. People were destitute and poverty stricken. The misery was increased during the war years because Turkey fought on the side of Germany and Austria against Britain and France, making it nearly impossible for funds from the Jews in the diaspora to reach the inhabitants of Eretz Yisrael. The death toll from hunger in the city of Jerusalem was enormous.

Rabbi Chaim Sonnefeld, the chief Rabbi of Jerusalem, had a plan to keep the inhabitants of Jerusalem from sinking into a state of despair. This story is about his love, kindness and caring, the qualities inherent in every "Yiddishe neshome" (Jewish soul)

Rabbi Sonnefeld's plan was to arrange matches between the young people of Jerusalem. He inquired of every person who came to him for advice if they were acquainted with any young people of marriageable age. When he found a young man and a young woman that seemed compatible, he immediately went to their respective parents to discuss the possibility of engagement. If the parents complained that they had no money for a wedding, a dowry or the price of an apartment, Rabbi Sonnefeld would say:

> "Contribute what you can towards the wedding feast, and He who is in charge of gold and silver will surely provide for the bride and groom."

So it was that amidst the despair in the Holy City, the "voice of gladness and the voice of joy, the voice of the bridegroom and the voice of the bride" resounded in Jerusalem.

Once, it happened that a "bride-to-be" sadly cried to her parents that she overheard her friends gossiping that her future husband was lame. Her father could not believe that Rabbi Sonnefeld would arrange such a match for his daughter so he went directly to him to demand an explanation. Rabbi Sonnefeld said he was acquainted with the young man, and he knew for a fact that he was not lame. He told the father of the bride not to worry.

As soon as the father left Rabbi Sonnefeld's house, the Rabbi went into his study, opened the box in which he kept his money and took out his last fifteen "lirot". He called his granddaughter Rivka to his side and said to her:

> "Take this money to the shoemaker located in the neighborhood of the groom's house. He is probably acquainted with the groom and must know his shoe size. Tell him to make a new pair of shoes for the groom. Then deliver the shoes to him. Tell the groom that the shoes are a wedding present from Rabbi Chaim Sonnefeld who recognizes his great diligence in the study of Torah."

The next day, Rabbi Sonnefeld went to see the worried father of the bride:

> "I have investigated the entire situation, and I am certain that your future son-in-law is not lame. I would like to invite you to see him for yourself. Please come with me to the 'yeshiva' where he is learning."

The bride's father went with Rabbi Sonnefeld and was very impressed with his future son-in-law. He begged forgiveness of Rabbi Sonnefeld for having succumbed to gossip.

Rabbi Sonnefeld had realized why it seemed as if the groom was limping. His shoes were so torn that he dragged his feet as he walked on the cobblestone streets of Jerusalem.

1 M. Getz. Yerushalayim Shel Maalah. Haginzar Hacharaydi. Jerusalem. 1977.

2 Jeremiah 7:34

"AHAVAT HABREYOT" DURING WARTIME[1]

Compassion, loving kindness, mercy: qualities which are inherent in "ahavat habreyot" describe the make-up of the Jewish personality. Rabbi Nathan bar Abba said in the name of Rav:

> "...whoever is merciful to his fellow men is certainly of the children of our father Abraham."[2]

There are no limits to acts of kindness which have the potential of promoting peace and understanding between people. Rabbi Eleazar ben Rabbi Yosi said:

> "All the charity and deeds of kindness which Israel performs in this world help to promote peace and good understanding..."[3]

It is with this in mind that Israel established an open border policy with southern Lebanon in the 1970's: the "good fence" was an important factor in the survival of the Christian forces as well as of Lebanese villages in the battle areas. Israeli health clinics treated thousands of villagers during the fighting. Often, the only source of food for many Lebanese was in Israeli markets.

The decade of skirmishes between Israel and Lebanon, culminating in the Israeli-Lebanese Peace for the Galilee War, will be researched and discussed by the historians of the future. However, during the war, the actions of Israeli citizens and soldiers were scrutinized by world opinion which tried to decipher if indeed they acted as fitting children of Abraham, imbued with "ahavat habreyot".

The following were reported in various Anglo-Jewish news journals:

Nursing students attending the Shapira Religious Vocational School spent their time in a club for the handicapped, replacing the nursing staff that had been called to military duty. (AMW Sept. 1982)

Students from all over Israel collected and packaged sweets, reading material, soap, insecticide, toothpaste and other necessities which were sent to the fighting men at the front.

Leaders of American Jewish organizations became actively involved in mobilizing its resources to raise emergency funds for Israel to replace the money used to aid the war effort which was earmarked for the normal functioning of humanitarian, cultural, welfare and educational services. (The Jewish Week, Sept. 10, 1982)

For the duration of the war, the mother of an Israeli soldier baked cookies in the early hours of the morning to bring cheer to her son's friends. Her son was stationed inside Lebanon and she knew he might never taste the cookies she baked. Yet, she baked and packaged the treats each day, then drove her car to the border, delivering them to the young men and women of the Israel Defense Forces.

Prior to entering battle, the Israeli troops were issued a document entitled "Image of An Israeli Soldier" (The Jewish Week, Sept. 3,

1982). It required no commentary. It demonstrated to the harsh critics of Israel that against the backdrop of the ugliness of war, facing a cruel enemy, Israeli soldiers were issued marching orders unique among armies. The document dared speak of "ahavat habreyot": treating the antagonist with kindness, the very fiber of Jewish tradition...

> "The Israel Defense Forces are now widely spread throughout Lebanon. Soldiers come in contact with large and varied civilian population. This contact places before you a challenge to personal behavior, Jewishly and militarily... It is forbidden to take booty, regardless of place... Do not harm the peaceful civilian population, especially the honor of women...
>
> "It is imperative for each IDF soldier to refrain from harming places which are cultural centers, antiquities, mosques, etc...
>
> "It is forbidden to harm holy places of all religions.
>
> "Every war raises feelings of hatred, jealousy, disrespect of life and property of the enemy. In spite of these feelings, each man must keep these ethical standards for they are the basis for the existence of human society.
>
> "IDF forces are fighting against a cruel enemy whose method of fighting employs terror against undefended citizenry. The IDF is forbidden to act in the same manner as is accepted by the enemy; it is forbidden to let feelings of revenge influence our behavior toward the Lebanese civilian population. Our unique strength derives from guarding the dignity of man."

One of the many beautiful stories about "ahavat habreyot" (The Jewish Week, Sept. 10, 1982) which was reported from the Lebanese frontier concerned a "Christian Phalangist" who turned out to be a Lebanese Jew. His family had made its way eastward four centuries before, at the time of the expulsion of the Jewish people from Spain, stopping in Turkey, enroute. Since Samu was a Lebanese citizen, he felt responsible to serve his country in the civil war between Moslems and Christians. Because Samu was a Jew, his army buddies called him "the Israeli" even though his roots in Lebanon are deeper than the roots of some of his comrades.

One day, Samu was walking around in a suburban area of Beirut when he heard Chassidic music blaring over a sound system. He rushed toward the sound of the music and recognized the Lubavitch Mitzva Tank which had been following the Israeli Defense Forces. Samu ran toward the music, shouting "Anoche Ivre, anoche Ivre," (I am a Jew, I am a Jew.) Coincidentally, the butt of Samu's rifle was decorated with the picture of Rebbe Menachem Mendel Schneerson, the Lubavitcher Rebbe, according to the custom of the Phalangist soldiers who decorate their rifle butts with pictures of their heroes.

Samu never denied his Jewishness, yet he never flaunted it publicly, since he was part of a minority in a country inhabited by Arabs, Moslems and Christians.

The Chassidim that manned the Mitzva Tank were overjoyed to find a Lebanese Jew. Samu was equally overjoyed to meet Israeli Jews and he immediately invited the Chassidim to visit his family.

As news of the encounter spread, other Lebanese Jews gathered at one of the houses to greet Israel's good will ambassadors in the true style of "ahavat habreyot". One of the many favors that the Lebanese Jews requested of the good will ambassadors was for them to arrange a fresh supply of kosher meat and poultry which had been cut off during the fighting.

The good will ambassadors also arranged for the other religious needs of the approximately 400 Jews that remained from an original population of 10,000 before the war. These needs included repairing the synagogue that had been damaged by the shelling. The Chassidim, themselves, brought the concrete blocks and tiles to repair the walls and floors and labored side by side with workmen from the Jewish community who volunteered their help.

The Mitzva Tank continues to follow the Israel Defense Forces. There will certainly be more stories about "ahavat habreyot" forthcoming from the many sources which report on Israel's involvement in Lebanon.

1 The stories in this section were gleaned from various Anglo-Jewish press releases reported in The Jerusalem Post, The Jewish Week, American Mizrachi Women, etc.

2 Talmud Bavli, Baytza 32 b

3 Talmud Bavli, Bava Batra 10 a

LOVE OF ISRAEL
אהבת ארץ ישׂראל

CHAPTER SIX

"Love for Eretz Yisrael is a fire burning in the Jewish heart."[1]

The prophet Ezekiel considered Eretz Yisrael to be "tabur ha-aretz" (the geographical center of the world) that connects three continents: Asia, Europe and Africa.[2] It has been coveted by the world's conquering powers since ancient times.

Eretz Yisrael is the land that G-d promised our Patriarchs as an inheritance for their children after them.

To Abraham, G-d said:

> "And I will give unto thee and unto thy descendants after thee, the land wherein thou art a stranger, all the land of Canaan for an everlasting possession."[3]

To Isaac, G-d said:

> "Live in this land and I will be with thee and bless thee, for unto thee and unto thy seed will I give all these countries, and I will keep the oath which I swore unto Abraham thy father."[4]

To Jacob, G-d said:

> "And the land which I gave unto Abraham and Isaac, to thee I will give it, and unto thy seed after thee, will I give this land."[5]

Moses, in his final blessing, delineated Eretz Yisrael as the land chosen by G-d to be the dominion of the Jewish people.

> "When the Most High gave the nations their place in the world,
> Fixing the borders of each one there;
> He took for himself the people of Israel,
> Jacob for his very own share...
> He gave him possession of the Promised Land."[6]

Joshua, the disciple of Moses, described Eretz Yisrael as "Eretz Chemdah", a land of yearning, of burning desire, a land without which the Jewish people could not live![7]

Since that time when Joshua conquered Eretz Yisrael, it has been the focal point of Jewish history. Its landmarks are simply starting points for exploring the past: Moriah, the mountain upon which the Holy Temple stood in Jerusalem, Meorat Hamachpayla, the burial sites of the Patriarchs and Matriarchs, Gerizim and Ayval, the mountains where the blessings and curses were pronounced, Midbar Ein Gedi, the desert where David confronted Saul, Carmel, where Eliyahu the prophet declared the Oneness of G-d, Usha, the home of the Sanhedrin, the Jewish court, after the destruction of the second Holy Temple. Through every inch of Eretz Yisrael, we can feel the reverberation of the soul of the land entwined within the soul of every Jew.

According to our sages, Eretz Yisrael has a special level of holiness, unfound in any other part of the world, attributed to G-d's Divine Presence which dwells in His Sanctuary (the Holy Temple in Jerusalem.)

Rabbi Elazar had such a high opinion of the holiness of Eretz Yisrael that he said:

> "Anyone who dwells in Eretz Yisrael lives without sin."[8]

From the time of both the Babylonian and Roman exiles, Jews have longed to return to Eretz Yisrael. The constant longing is reiterated in our daily prayers for G-d to rebuild Zion.

There was always a spiritual connection between the Jews who lived in the exile and the Jews of Eretz Yisrael. Jews born in the exile went to Eretz Yisrael to study Torah in its great academies. Wealthier Jews provided generous financial assistance for the upkeep of the second Holy Temple while it stood in Jerusalem. After its destruction, the charity money was used to provide sustenance for the needy, a custom which has continued for the past twenty-five hundred years. Jews living in exile left instructions in their wills to be buried in Eretz Yisrael because Eretz Yisrael is called "Eretz Hachayim", the land of eternal life.[9]

The dream of Jews all over the world is to see the fulfillment of the vision:

> "And I will bring you to the land which I promised to Abraham, Isaac, and Jacob, and I will give it to you as an heritage."[10]

This dream was beautifully described by Rebbe Nachum of Bratslav when he returned to Russia from pilgrimage to Eretz Yisrael in the early part of the nineteenth century. He said:

> "Bechol makom sheani holaych, ani holaych l'eretz yisrael. (Every place I go, I am going to Eretz Yisrael.)"[11]

The stories in this chapter are about Jewish yearning for Eretz Yisrael.

1 Rebbe Shneor Zalman of Liadi. Tanya. Slavita. 1796.
2 Ezekiel 38:12
3 Genesis 17:8
4 Genesis 26:3
5 Genesis 35:12
6 Deuteronomy 32:8, 13
7 Nodeh Lecha, Birkat Hamazon.
8 Talmud Bavli, Ketuvot 111 a
9 Ezekiel 26:20
10 Exodus 6:8
11 Bayt Hatefutzot: The Museum of the Diaspora, Tel Aviv.

YEARNING FOR ZION[1]

Rabbi Yehuda HaLayvi was the poet laureate of the Golden Age of Spain, that period of Jewish history in which mutual coexistence between Moslem and Jew existed for more than three centuries. He lived in Cordova, one of the most prominent cities in Moslem ruled Europe. Science, literature and medicine flourished in Cordova. The city had a great library, nearly half a million volumes, and since paper was gradually being substituted for parchment and books were more easily accessible, it became an attraction for scholars to live and study in this great center where Judaic learning and general Islamic culture flourished side by side.

Rabbi Yehuda HaLayvi was an only child and his family made certain that he was well educated. He studied Torah with one of the great codifiers of Jewish law, Rabbi Yitzchak Alfasi, in addition to acquiring the knowledge of medicine, philosophy and Arabic.

Using the Hebrew language as a vehicle for its poetic expression, he composed more than three hundred verses and elegies, many of which are included in the Selichot (penitential prayers), recited before Rosh Hashana, and the lamentations (mourning dirges), describing the destruction of the Holy Temple in Jerusalem, recited on Tisha B'Av.

Rabbi Yehuda HaLayvi created poetry that seared the soul of his readers. His themes seem to be drawn from Divine inspiration. Some critics compare him to King David. As King David was called "Neim Zemirot Yisrael"[2] (the Sweet Singer of Israel), the composer of the "Tehillim" (Psalms), so Rabbi Yehuda HaLayvi was called the "Singer of Zion", composer of an emotionally filled liturgy reflecting his yearning for Eretz Yisrael.

The major themes of these liturgical and poetic verses describe the suffering of the Jewish people in the Diaspora. He anguished over the fact that the only memorial to the glory of ancient Jewish history was the remnant of the Holy Temple, the Kotel, the Western Wall. His poetry is full of beautiful prayers expressing the hope that the "Shechina" (G-d's Holiness) would return to Zion as in the days of old.

The following are two examples of the deeply religious poems which express the deep love of Rabbi Yehuda HaLayvi for Zion. They are recited on Tisha B'Av, the national day of Jewish mourning in memory of the destruction of both Holy Temples.

ZION, WILT THOU NOT ASK

"Zion, wilt thou not ask if peace be with thy captives
That seek thy peace - that are the remnant of thy flocks?
From west and east, from north and south the greeting
Peace from far and near, take thou from every side
And greeting from the captive of desire, giving his tears like dew
Of Hermon, and longing to let them fall upon thine
hills..."[3]

MY HEART IS IN THE EAST

"My heart is in the east, and I in the uttermost west
How can I find savour in food? How shall it be sweet for me?
How shall I render my vows and my bonds, while yet
Zion lieth beneath the fetter of Edom, and I in Arab chains
A light thing would it seem to me to leave all the good things of Spain,
Seeing how precious in mine eyes to behold the dust of the desolate sanctuary."[4]

All his life, Rabbi Yehuda HaLayvi yearned to step on the soil of Eretz Yisrael where he could pour out his soul at the place where the Divine Presence dwells. He dreamed of settling in Eretz Yisrael so that he might live a spiritual life in holy surroundings.

After spending the major part of his life in exile, Rabbi Yehuda HaLayvi decided to undertake the very hazardous trip to Eretz Yisrael. He was sixty years old. The boat upon which he sailed the Mediterranean was pounded by violent windstorms and the sailors feared for their lives. Rabbi Yehuda HaLayvi was so confident that his yearning for Eretz Yisrael would culminate in his standing at the Western Wall in prayer, that he remained calm throughout the voyage.

According to this version of the story[5], as soon as the boat docked in Eretz Yisrael, Rabbi Yehuda HaLayvi started his pilgrimage up toward Jerusalem. When his eyes beheld the gates of Jerusalem, a vision that had filled him with burning desire for years, he became very emotional; and upon witnessing the desolation of Jerusalem, he tore his clothes and cried. He bent down to kiss the holy soil, and the words of "Zion, Wilt Thou Not Ask" blurted forth from the depth of his soul. Then he proceeded to the "Kotel", the Western Wall.

He had finally achieved his dreams! Rabbi Yehuda HaLayvi wrapped himself in "talit" and "tefillin" (prayer shawl and phylacteries), his hands outstretched, touching the stones of the Holy Wall, his eyes streaming with tears. As he was praying, an Arab horseman, seeing a Jew expressing such intense love for Eretz Yisrael, trampled him. He died in front of the Western Wall. He was buried in the village of Kabul, in the territory allotted to the tribe of Asher, in the Galilee.

1 J.D. Eisenstein. _Otzar Yisrael._ Shiloh Publishing Co. Jerusalem.
2 _II Samuel._ 23:1
3 _Heinrich_ Brody, editor. _Selected Poems of Yehuda Halayvi._ The Jewish Publication Society. Philadelphia. 1974.
4 _Ibid._
5 _Op. cit._ _Otzar Yisrael._

AN ANCESTOR HELPS AVERT DOOM[1]

The Moslem Ottoman Turks ruled the Middle East from the fifteenth to the early twentieth centuries. Islamic tradition dictated religious tolerance for people who believed in monotheism, so the Jews were respected and well treated throughout the vast and sprawling empire. In fact, Sultan Bayazio mocked the Spanish King Ferdinand, of Inquisition notoriety, when he said:

> "Ferdinand expelled the Jews to impoverish his country in order to enrich mine."[2]

Eretz Yisrael was under the sovereignty of the Ottoman Turks. Many Jews who fled from religious persecution in Spain and other Christian countries in western Europe immigrated to Eretz Yisrael, hoping to restructure their lives in the Holy Land.

They chose Eretz Yisrael because it was "home". It was the land which G-d promised to the patriarchs and matriarchs, the land saturated with religious and historical significance, the land which our people had never completely abandoned in all the years since the destruction of the second Holy Temple and the consequent exile.

Most of the Jews that lived in Eretz Yisrael during these centuries settled in the Galil region, the northern part of the country, where commerce flourished along the trade routes that connected the continents. There was also a Jewish community in Jerusalem. The Jerusalem community was comprised of Jews who believed it to be the spiritual center of the world.

This story is about the Rebbetzen (wife) of Rebbe Shmuel Kaminka, a Chassidic master of the third generation following the Baal Shem Tov, who longed all his life to make "aliya". Because he was so poor, his plans had never come to fruition.

The village of Kaminka had two semi-annual fairs, one after Pesach (Passover) and the other after Succot (Tabernacles). Once during a fair, a merchant knocked on the door of Rebbe Shmuel Kaminka's house. He was carrying a heavy package. When the Rebbetzen opened the door, the merchant asked if he could leave his package for safekeeping in her home since it was too heavy to carry around all day while he conducted his business at the fair. He promised to return that evening to pick it up. She agreed and stored the package under the bed, not bothering to tell her husband about the incident.

The merchant did not return that evening and the hidden package was soon forgotten. Imagine the surprise of the Rebbetzen when she began cleaning her home in preparation for the following Pesach. While cleaning underneath the bed, making certain there was no leaven any place in her house, she retrieved the forgotten package. She recalled the incident and disclosed it to her husband. They opened the package and found that it was filled with gold coins.

Rebbe Shmuel Kaminka said immediately:

> "This gift from heaven will enable us to make 'aliya' (to go up to Eretz Yisrael and settle there)."

Immediately following that Pesach, they left Kaminka for Eretz Yisrael. This was about the end of the first decade of the nineteenth century.

Rebbetzen Kaminka was a very righteous woman who was reputed for being an outstanding midwife; she could miraculously deliver babies instantaneously and painlessly. Her reputation followed her to Eretz Yisrael.

One day, Rebbetzen Kaminka was called on to assist in the birth of a child being born to the daughter of the Ottoman Turkish governor of the Jerusalem area. The governor pleaded with the Rebbetzen to try to save his daughter's endangered life and that of the unborn child. Having consented to help, she entered the room where the governor's daughter was laboring and made her comfortable. Shortly thereafter, the child was born. The governor was overjoyed with the birth of his grandchild and without hesitation asked her to name her reward.

This righteous woman seriously thought over all the possibilities about which she dreamed and knew that she might never attain. She decided on the one thing that she dreamed about most: to be safely accompanied to Meorat Hamachpayla, the gravesites of the Patriarchs and Matriarchs in Hebron, a place that was prohibited to Jews. She told him of her desire to visit the gravesite in three months, knowing how much time she needed to prepare herself through fasting and prayer. The governor granted her request.

When the governor's grandchild was three months old, he sent two soldiers to accompany Rebbetzen Kaminka from Jerusalem to Hebron. The governor was unaware that the soldiers he chose hated Jews. The soldiers knew they had to follow the governor's orders to accompany this righteous woman safely to Hebron. Since the governor did not instruct them to bring her back to Jerusalem safely, they decided they would leave her at the gravesite of the Patriarchs and Matriarchs, burying her with her ancestors.

This righteous woman, unaware of the soldiers' evil plans, eagerly anticipated the fulfillment of her dreams. Her emotions and joy, upon her arrival at the gravesites, were indescribable. The soldiers helped her descend to the underground caverns where the actual graves were located. She stood in prayer over the graves of Abraham and Sarah, then walked to the graves of Isaac and Rebecca, and then to those of Jacob and Leah.

When she finished praying, she signalled to the soldiers that they might help her ascend to ground level. There was no answer. At that, she walked over to a corner, sat down and began to recite the Psalms that she remembered from memory. After a long while, the joy she felt at the fulfillment of her dream turned to anguish. She wondered if she would ever be reunited with her family or if she would ever deliver another baby.

Suddenly, from a dark corner of the gravesite, she noticed a faint light. She was barely able to distinguish the light of a candle being held by a short man. As the man drew near, the light grew stronger and more intense. She saw that the man had red curly hair and a red beard. The man took her by the hand and urged her to follow him and together they ascended to the outside of the gravesite. Then, she found herself standing on the doorstep of her home in Jerusalem, alone. In her hand, she held a book of Psalms which she realized had been given to her by its author, King David, "Neim Zemirot Yisrael", the sweet singer of Israel.[3] Tradition has it that the book of Psalms was handed down to her children, and they to theirs for many generations. It remained in that family until it disappeared during an Arab skirmish in the old city of Jerusalem during the second World War.

1 Sh'moneh Seforim. Bayt Pinchas. Bulgaria. 1913. Reprinted in New York in 1983. This book is an anthology of Torah discourses and stories. There is no single author.

2 Encyclopedia Judaica. Volume 16. p. 1533.

3 II Samuel 23:1

YOU COULD WEAR THESE PANTS EVEN IN JERUSALEM[1]

The eighteenth and nineteenth centuries were filled with anguish and uncertainty for eastern European Jewry. Sovereignty over the Polish people had shifted as Russia, Austria and Prussia thrice partitioned the Polish empire. The Pale of Settlement, that vast geographical expanse, about one million square miles between the Black and Baltic Seas, had a very large Jewish population. This area was gobbled up by Russia.

The religion of the Russian people was Russian Orthodox, a branch of the Byzantine Catholic Church. The heads of the Church resided in the Moscow area and they were very suspicious of foreigners. They did not tolerate strangers dwelling within Russia's borders.

When Tzarina Catherine the Great annexed the Pale of Settlement, she realized she had a tremendous problem, namely, how to accommodate Polish Jews who were living under Russian rule. She solved the problem by confining the Jews to the Pale of Settlement. Succeeding Russian Tzars tried to deal with the Jewish problem by eliminating the Jews altogether. Tzar Nicholas I set up a diabolical plan to hasten Jewish assimilation through forced military conscription for twenty-five years. Tzar Alexander III devised a plan to forcibly convert one third of the Jews, force one third to immigrate, and methodically murder the rest.

As a result of this persecution, the Jewish national spirit awakened, and the impetus to return to Eretz Yisrael began to grow. Among the religious leaders who spoke out to rebuild Eretz Yisrael was Rabbi Tzvi Hirsh Kalischer. Rabbi Kalischer and his contemporaries agreed that Eretz Yisrael was the land of the Jewish future and it was incumbent upon the Jews to begin the work of returning home. He believed that the Jews living in western Europe had the wealth and the political influence to help their brothers in Russia. He proposed to use the financial aid of western European Jews to establish agricultural settlements in the Holy Land and the labor of Russian Jewish immigrants to maintain them. He aspired to make the desolate waste that lay fallow for the past two thousand years blossom and flourish again.

About the same time, a group of Jewish university students, inspired by nationalistic fervor, were convinced that the time to end the exile mentality of the Jew, had arrived. These young Jews called themselves "Biluim", an acronym for :"Bayt Yaakov, Lechu V'naylcha" (House of Jacob, let us get up and go).[2] The Biluim rejected immigration to any other country because they believed that Jews who continued to wander from country to country would only extend the exile even though they may have found temporary happiness and security outside of Eretz Yisrael. The Biluim wanted to encourage Jews to return to the land of their roots, the land which gave birth to the Jewish nation.

The Biluim used as their symbol the six pointed Star of David, inscribed in each corner with one of these words:

"Hakatan yeheh l'elef, v'hatzair l'goy atzum" (the smallest will become thousands and the youngest will become a great nation).[3]

In the center of the star, two hands grappled, holding fast.

When the Biluim reached the Holy Land, they laid the foundation for the development of agriculture, industry and military training.

The yearning to make "aliya" was very strong throughout this period of upheaval and persecution. This story is about one Jew who yearned to make "aliya", but he did not immigrate until a sign was revealed to him.

He was a Jew of substance whose entire livelihood depended upon his relationship with the "poretz" (the noble landowner) of his village. The Jew had a beautiful daughter of marriageable age named Sarah. She was G-d fearing, intelligent, and kind.

The "poretz" had a son who noticed the girl when she came to the village marketplace. He tried repeatedly to gain her attention but she ignored him continuously. The more she ignored him, the more he persisted to the point that he imagined he was madly in love with her and decided that he would marry her.

Sarah was alarmed by the young man's persistent nagging, so she told her father. He realized that the situation could get out of hand. He knew that the only way to save his daughter from the "poretz's" son was to marry her off to a Jew as soon as possible. He did not think the "poretz's" son would continue paying attention to his daughter if he knew she were married.

The only unmarried Jewish man in the village was the tutor who earned his livelihood by teaching the children in the village. He was a kind, scholarly man, but much older than Sarah. He lived in Sarah's father's house. Sarah's father asked him if he would consent to marry his daughter. The tutor agreed on the condition that if the marriage did not work out after a year, he would willingly grant her a divorce. Sarah agreed to the match in order to be freed from the unwelcome attentions of the "poretz's" son. Sarah's father arranged the wedding of his daughter to the old tutor.

After a year, the old tutor wanted to fulfill the conditions of the marriage, so he volunteered to divorce his wife. Sarah refused to accept the divorce. Instead, she said to her husband:

> "I pray that G-d will bless us with a son who will follow in the righteous footsteps of his father."

G-d answered her prayer and Sarah gave birth to a son. She named him Leib, but throughout his lifetime he was known by the name of Leib ben Sarah (Leib, the son of Sarah), in honor of his mother who married the old tutor to escape from the "poretz's" son. When he grew up, Leib ben Sarah became a great Chassidic Rebbe. He was called Rebbe Leib Sarahs.

It was his custom to collect charity money and distribute it to the poor. One morning, he rushed breathlessly into the synagogue. A few people gathered around him to see why he was breathless and he explained that he needed to borrow three hundred rubles desperately. The people knew that he was a righteous man and they trusted him.

A business man stepped forth from the group and said:

> "I happen to have three hundred rubles that I was about to invest in a business. I will give it to you."

Rebbe Leib Sarahs said:

> "You don't have to worry about your money. I bless you that you will have much more than you are lending me today."

When the business man saw that Rebbe Leib Sarahs blessed him, he said:

> "Please promise me that since I lent you this money, you will introduce me to one of the 'Lamed Vav Tzadikkim'."[4]

Rebbe Leib Sarahs agreed to introduce the businessman to a "Lamed Vav Tzadik".

Many months passed and Rebbe Leib Sarahs blessing was fulfilled. The businessman prospered more than he could possibly dream. He accumulated so much money that he began thinking about retiring and settling in Eretz Yisrael to live the remainder of his life. He had been yearning for the privilege of living in the Holy Land for many years.

He wanted to ask Rebbe Leib Sarahs' advice, so he went to him and said:

> "I want to know if I should go to live in Eretz Yisrael, and furthermore, you haven't introduced me yet to a 'Lamed Vav Tzadik'?"

Rebbe Leib Sarahs answered:

> "Next week, after the Sabbath, around midnight, go to the little hut that you will see as you drive your wagon out of the village. It is a tailor's hut. The tailor will advise you what to do with the remainder of your life."

The businessman followed Rebbe Leib Sarah's advice. After the next Sabbath, he harnessed his horse, hitched up his wagon, and took off for the hut at the outskirts of the village. He wanted to knock on the door and go into the hut, but he was embarrassed to enter without an excuse, so he tore his pants and then he knocked on the door. The tailor opened the door, welcomed him warmly and asked how he could help. Noticing the torn pants, he took them from the businessman's hand, laid them on his sewing counter and began to sew.

With each stitch, the businessman squirmed uncomfortably. As the tailor pierced the fabric with his needle, the businessman felt as if his soul was being pierced. As the tailor pulled the needle through the fabric, with each consecutive stitch, the businessman sensed that his soul was rejuvenated. When the tailor completed sewing the pants, the businessman perceived that his soul had been redeemed. He was aware that the tailor had restored his soul, but he still did not know if he should make plans to go to live in Eretz Yisrael.

While the businessman was standing in the tailor's hut reflecting upon the situation, Rebbe Mottele Tzernobler, a disciple of the Baal Shem Tov, entered. He brought a pair of pants for the tailor to mend. The tailor took the pants from Rebbe Motele's hands and began to sew them. With each stitch of the tailor's needle and thread, Rebbe Motele squirmed. He cried with pain when the needle pierced the fabric, and sighed with joy when the tailor pulled the needle through. The tailor was restoring Rebbe Motele's soul.

The tailor was almost finished mending the pants when he turned to Rebbe Motele and said:

> "I mended these pants so well that you could wear them even in Jerusalem."

The businessman stood in awe of what went on in the tailor's hut. He understood that the tailor had not only restored his soul but he had also answered his question of whether to go to Eretz Yisrael. He realized that Rebbe Leib Sarahs had directed him to one of the "Lamed Vav Tzadikkim".

The businessman made immediate plans to make "aliya", to live the remainder of his life in Eretz Yisrael.

1 Op. Cit. p. 136. Shmoneh Seforim.
2 Isaiah 2:5
3 Isaiah 60:22
4 Op. Cit. p. 50. Talmud Bavli Succah 45 b.

IF YOU WILL IT, IT IS NO DREAM[1]

Parallel with the emergence of Zionism in eastern Europe, Jewish nationalism made an appearance in western Europe. The above movement was stimulated by an anti-semitic incident.

The trial of Captain Alfred Dreyfus, which shook all of France to its core, made Jews painfully aware of the fact that they were strangers in a foreign land.

Jews had beem granted equal rights in France by Napoleon Bonaparte at the beginning of the nineteenth century. They were free to adopt the western way of life which had so long been closed to them by persecution and ghetto isolation.

A period of approximately one hundred years passed between Napoleon and Dreyfus. During that time, many French Jews assimilated into French civil society. These Jews considered themselves French citizens and some attained high government office.

This story is about a French born Jew, an army captain, named Alfred Dreyfus, who was accused of treason against his native country for selling war secrets to the Germans. One result of this infamous trial was the founding of political Zionism.

During the lengthy trial, it became obvious that Captain Alfred Dreyfus was framed because he was born of the Jewish faith. The crowd at the trial yelled vociferously:

"Death to the traitor, death to the Jew!"

Although Dreyfus steadfastly declared his innocence, French army leaders insisted that he was not worthy to be a French soldier. Captain Dreyfus was found guilty by the French court. His rank stripes were torn from his jacket; he was publicly disgraced and sentenced to exile.

Emile Zola, an influential French novelist spoke out against the shameful verdict because he believed that Dreyfus was innocent. He accused the judges of prejudice and collusion. Zola demanded impartiality. He wrote a letter "I Accuse" to the French president in which he demanded to know how the enlightened president of an enlightened country could permit an innocent man to be framed during his regime. Zola was condemned for libeling the French government and was subject to harassment. To escape imprisonment, he fled to England. His accusations caused such an uproar that they divided French citizens into pro-Dreyfus and anti-Dreyfus camps.

Finally, after three tumultuous years, Dreyfus was granted a new trial that was equally marred by prejudice. New evidence which was introduced to prove Dreyfus' innocence was squelched by French army leaders. The generals of the French army still believed that the honor of the army was more important than the life of a Jew. Dreyfus was sentenced to ten years in exile, but the French president pardoned him. The proof of his innocence was not established until years later when France's highest court reviewed the facts and declared him innocent. At that time, he was restored to his rank in the French army and the perpetrators of the injustice were duly punished.

Theodore Herzl covered the trial of Captain Alfred Dreyfus. He was an assimilated Austrian Jewish journalist and his newspaper, "Neue Freie Presse", had sent him to Paris to cover the trial. When he heard the frenzied mob screaming, "death to the traitor, death to the Jew," he realized that no matter how much equality Jews attained, it would be impossible to erase social hostility. He learned from previous travels that the rights granted to Jews were different in every country where Jews lived. Herzl believed that Jews would never find peace in "galut" (exile). He became obsessed with the idea of a national homeland for the Jewish people, and this obsession plagued him day and night. He published a prophetic but controversial book, The Jewish State. In it, Herzl presented a practical program for the future settlement of the Jewish people in their land with the consent of the great world powers. He wrote of Jewish survival among the nations of the world since the destruction of the Second Holy Temple and that the time had arrived for a return to their national homeland. He proposed that the state be supported by immigrant workers and financiers and it would be the catalyst for restoring the national life of the Jewish people. Herzl suggested that the problem of Jews living in foreign countries would not be solved by assimilation because the Jews did not want to disappear into a non-Jewish world.

Many Jews agreed with Herzl and supported his ideas. A World Zionist Congress was organized and held its first convention in Basel, Switzerland, in August, 1897.

Then Herzl began travelling to see various heads of state in order to convince them to support his idea of a Jewish homeland. The Turkish Sultan agreed to an audience with Herzl, but insisted that Jews could live anywhere in Turkish ruled lands but they could not concentrate in Eretz Yisrael.

Herzl travelled to see the Pope, and had audiences with members of the Italian government, Russian ministers and English statesmen. He knew that the German Kaiser, Wilhelm II, was planning to visit Jerusalem, so Herzl made an appointment to meet the Kaiser in a tent camp outside of Jerusalem in the fall of 1898. The Kaiser listened patiently to Herzl's plea, but when the Kaiser returned to Germany, it became apparent that he had no interest in helping Herzl succeed. He was invited to present his plea for a Jewish homeland before the English rulers. England responded by appointing an investigative commission to check the feasibility of establishing a Jewish homeland in the Sinai peninsula, an area under its rule. The commission abandoned the idea because the Sinai peninsula lacked water. Instead, they proposed the African territory of Uganda as a Jewish homeland, but the world's Zionist leaders rejected this proposal because they viewed Uganda as another place of exile like any other country and not the birthplace of the Jewish nation. The Zionists insisted on claiming the right to settle in Eretz Yisrael based on the verse:

> "If I forget you, Jerusalem, may I forget the use of my right hand."[2]

Herzl pushed for the fulfillment of his dream even though his physical condition was weakened by a frail heart. He wrote another book <u>Alt Neu Land</u> which is an idealistic vision of the way the new state would function in the future. Herzl's dream for the establishment of the new land was based on scientific technology, agricultural development, industrialization, progressive government, cultural life, and toleration between Arab and Jew.

Herzl did not live to see the fulfillment of his dream. His heart gave out at the age of forty-four. He died in Vienna and was buried there. With the establishment of the State of Israel, his body was moved and interred on Mt. Herzl in Jerusalem. His dream became the motto of the Zionist movement:

> "Im tirtzu, ayn zo agada"
> (If you will it, it is no dream.)

1 <u>Op. Cit</u>. p. 133. <u>Otzar Yisrael</u>.
2 <u>Psalms</u> 137:5.

L'CHAYIM[1]

The devastation of the Holocaust plagued the conscience of the world and influenced the leaders of some nations to support the establishment of an independent Jewish state in Eretz Yisrael. The British, who held a mandate over Eretz Yisrael, could not accept the suggestion of an independent Jewish state because they were unable to reconcile the differences between Arabs and Jews. The British turned the problem over to the fledgling United Nations for solution.

At a United Nations session in April, 1947, a special committee on Palestine (UNSCOP) was appointed to investigate the feasibility of a Jewish state, an Arab state and a small internationally administered zone to include Jerusalem. The resolution for partition was presented to the General Assembly of the United Nations which adopted the recommendations in November, 1947. The Arabs rejected the partition plan. The Jews, though unhappy with the plan, accepted it.

The British began to withdraw from Eretz Yisrael early in 1948. When the British High Commissioner for Palestine departed on May 14, 1948, David Ben Gurion, head of the Provisional Council of State convened the council to approve the Declaration of Independence for the new State of Israel.

The blessing "shehecheyanu, v'keymanu, v'hegeyanu lazman hazeh" (Who has kept us alive and preserved us and enabled us to reach this season) was recited and the Declaration of Independence was unanimously adopted. The State of Israel was born.

With the birth of the Jewish state, a majestic sunlight had dawned. The dreams and yearnings of generations, the innermost longings of the Jewish soul slowly began to be realized.

The tragic words of the prophet Jeremiah's lament over the destruction of the Holy Temple:

> "For these things I weep, my eyes, my eyes are running with tears, because the comforter is far from me and there is no one to restore my soul,"[2]

were changing into the words of comfort and hope prophesied by Isaiah, the comforter of the Jewish people:

> "Be comforted, be comforted My people...Do not be afraid, because I have returned you to your own land."[3]

The following story is about a Chassidic survivor of the Holocaust who managed to make his way to Eretz Yisrael after World War II. The story depicts the intense love and devotion that this man had for Eretz Yisrael. It provided him with the faith to withstand his own personal suffering.

The man was in Eretz Yisrael a very short time when the War of Independence broke out, and he volunteered to serve in the defense of the Jewish state. After the War of Independence, he thought there would be no more war, but, years later, he was called upon again to defend his country during the Sinai campaign.

The man had two sons, both born in Israel, and by the time the Six Day War broke out, both were old enough to serve in the army. The oldest son was killed during the Six Day War. The youngest son was really not required to serve, but he insisted on the privilege of fighting during the Yom Kippur War, six years later. The toll of young Israelis who fell in battle during the Yom Kippur War was enormous. The second son of the Chassidic survivor was among the dead.

The dead boy's lieutenant commander had the task of breaking the tragic news to the parents. He didn't have the courage to face them. He drove to their apartment, parked his car and walked around the block seven times before he summoned enough strength to knock on their door.

The father welcomed the lieutenant commander to the apartment. As they sat down, the lieutenant commander asked that the man call his wife into the room. When she entered, the lieutenant commander said in a low voice:

> "I have come to tell you," he said, "that your second son was slain in battle."

Upon receiving this tragic news, the man rose, walked over to a cabinet, took out a bottle of wine with two glasses, poured the wine into each glass, gave one to his wife and kept one for himself. He raised his glass and said:

> "L'Chayim, Master of the world. I have given you two holy sons, one in the Six Day War and one in the Yom Kippur War. L'Chayim, my wife. I gave you the privilege of bearing two holy sons who died in the defense of the Holy Land. L'Chayim, L'Chayim."

With this, he shook the hand of the lieutenant commander and said:

> "Baruch dayan ha-emet."
> (Praised be the Eternal, Master of all Truth.)

1 Rabbi Shlomo Carlebach told this story to me soon after he returned from Eretz Yisrael following the Yom Kippur War. He flew to Eretz Yisrael to bring moral support to Jewish soldiers who fought with the Israel Defense Forces.

2 Aycha Lamentations 1:16.

3 Isaiah 40:1, 43:3.

A MESSAGE FROM THE KOTEL[1]

This story poignantly describes what many have believed for thousands of years, namely, that the "Kotel" is like a bridge which fills the void between heaven and earth. From the "Kotel", our prayers ascend aloft to heaven. Our Rabbis explained that no prayer recited at the "Kotel" goes unheeded. They based their explanation on a statement of the sages which says that the "Kotel" has ears.[2] The "Kvittels" (the little pieces of paper inscribed with prayers), tucked into the crevices of the "Kotel's" stones bear witness to the belief that our prayers are heard at the "Kotel". The "Kotel" is also the place where the love of Eretz Yisrael is expressed from the depths of the Jewish heart.

A soldier approached Shlomo and said:

> "Please follow me. I came to return a 'kvittel' to the place in the Holy Wall from where I took it. Let me tell you why I took it out.
>
> "My father wrote a 'kvittel' two days before he died and placed it among the thousands that are tucked into the Holy Wall. The 'kvittel' consisted of a prayer that G-d should protect me no matter what events I encountered.
>
> "My father passed away last summer, and when I was drafted, I ran to the Holy Wall to say kaddish one last time before I joined my comrades. I wanted to put a new 'kvittel' in the Holy Wall, but it was the Sabbath and I couldn't write, so instead I pulled down an old 'kvittel' to take with me. When I opened the 'kvittel' that I had pulled down, I saw that it was the one my father wrote for me. I took the 'kvittel' with me to the battlefield. I believe that my father's prayer was answered and I survived.
>
> "Tonight, I have come to return the 'kvittel' to the Holy Wall."

1 The following story is a true experience which happened to Rabbi Shlomo Carlebach as he was preparing to welcome the Sabbath one Friday night at the "Kotel" (the Western Wall).

2 Vayikra Raba 32.

MESSIAH
מָשִׁיחַ

CHAPTER SEVEN

"The Messiah will not appear until redemption has been despaired of, when there seems to be no supporter and helper for Israel."[1]

The exact description of the Messianic Age has not been revealed to man,[2] so our sages have attempted to depict what will occur in the end of days. They base their opinions on various verses from Scripture and we must depend on their imaginative interpretations for clues.

> "And it shall come to pass in that day,
> That the Lord will set His hand again the second time
> To recover the remnant of His people...
> And He will give the signal to the nations
> And will assemble the dispersed of Israel,
> And gather together the scattered of Judah
> From the four corners of the earth."[3]

Our Rabbis interpret the above verse to mean that the Jewish people will be redeemed from exile and they will be gathered into their own land in the Messianic Age. Likewise, they interpret the following verse to mean that the Messianic Age will bring peace to the world and all mankind will recognize the glory of G-d.

> "For the earth shall be full of the knowledge of the Lord, As the waters cover the sea."[4]

Unlike the Messianic concept in Christian doctrine, the Jewish redeemer will not have to offer his life to atone for the sins of mankind. Rather, his throne will be the foundation of justice. He will be endowed with the instinct to decipher right from wrong. He will be a prophet, warrior, judge, king and the teacher of Torah. He will restore the Jewish people to their land and direct the rebuilding of the Holy Temple in Jerusalem.

From there, the Kingdom of G-d will be proclaimed.

> "And there shall come forth a descendant from the stock of Jesse,
> And offspring from his roots shall flourish
> And the spirit of wisdom and understanding shall rest upon him,
> The spirit of counsel and strength
> The spirit of knowledge and of the fear of the Lord.
> And his delight shall be in the fear of the Lord,
> And he shall not judge by appearance,
> Neither decide by hearsay,
> But with righteousness shall he judge the helpless,
> And decide fairly for the humble of the land,
> And he shall strike down the ruthless with his verdicts
> And with his sentences.
> And justice shall gird him up for action
> And He shall be belted with trustworthiness."[5]

Our Rabbis have infused our imaginations with verbal descriptions of the Messianic Age. One verbal description depicts the generation preceding the coming of the Messiah:

> "Scholars will be few in number,
> There will be great pain like the travail of childbirth,
> Evil decrees will be promulgated,
> Tragedy, war, each evil arriving hastily on the heels of
> The preceding evil."[6]

It concludes with the promise that:

> "Three days before the Messiah, Eliyahu the prophet of peace and redemption will appear and proclaim: 'Mountains of Israel! How long will you remain desolate and waste? Peace has come to the world. This is the coming of the age of redemption. This is the beginning of the great day.'"[7]

All the above descriptions are enigmatic. The only certainty has been the profound yearning for redemption when Jewish suffering increased. During the periods of Roman persecution, the wanderings and expulsions of the Middle Ages, the Crusades, the Inquisition and the Holocaust, the quest for redemption was more intense than during periods when Jews lived in relative security.

According to the Midrash, the Messiah was born on Tisha B'Av, the day the Holy Temple was destroyed.[8] The Messiah is circulating somewhere in this world, waiting for the moment of redemption. Maybe the Messiah will be revealed on Tisha B'Av. Maybe the reestablishment of a Jewish homeland on our ancestral soil is the "Atchalta D'geula" (the beginning of redemption). Until the Messiah arrives, we continue to believe in his coming. This belief is one of the fundamental principles of the Jewish faith.

> "I believe with perfect faith in the coming of the Messiah, and though he tarry, I will wait daily for his coming and never lose faith."[9]

The stories in this chapter reflect the yearning of the Jewish people for the Messiah.

1 Talmud Bavli Sanhedrin 97 a
2 Talmud Pesachim 54 b
3 Isaiah 11:11,12
4 Isaiah 11:9
5 Isaiah 11:1-5
6 Talmud Sanhedrin 97 a
7 Talmud Pesachim 88 a
8 Aycha Raba 1
9 Maimonides. Shlosha Asar Ikarim (The Thirteen Principles of Faith)

RABBI AKIVA'S FAITH IN THE FUTURE REDEMPTION[1]

As the years following the destruction of the Second Holy Temple lengthened into generations, discussions of redemption persisted among the leaders of the Jewish people. They believed that the Messiah would follow on the heels of great Jewish suffering. There had been no more greater tragedy in Jewish life than the destruction of the Second Holy Temple and the loss of Jewish independence.

The Romans were determined to squelch any hopes of Jewish spiritual survival. They hoped that by instituting the same restrictions which the Greek-Syrians had imposed four centuries before, they could destroy the Jewish nation entirely. The Romans prohibited the observance of the Sabbath, the rite of circumcision, the sanctification of the new moon, adherence to Kashrut, and the study of Torah.

The Jews that were threatened by Roman oppression reacted the same way their ancestors did during the Greek-Syrian oppression four centuries before. They challenged the might of the Roman empire, hoping to surmount the oppression and triumph over the hated foreigners who were imposing their iron will on them.

The Romans underestimated the power of the Jewish people to survive. During the period of the Greek-Syrian decrees, the Maccabbees rebelled and freed Eretz Yisrael from tyranny. The leaders of the Jewish people decided to follow the example of their Maccabean ancestors. First, they fearlessly defied the Roman decree against the study of Torah. Then, they began to encourage the young men to organize a rebellion against Rome.

Rabbi Akiva, whose name was synonymous with the fight for Jewish independence, was in the forefront of those leaders who felt that the time had come to throw off the yoke of Roman rule and rid Eretz Yisrael of the hated foreign master. He believed that the arrival of the Messiah was imminent.

One day, Rabban Gamliel, Rabbi Eleazer ben Azariah, Rabbi Akiva and Rabbi Yehoshua ben Chananya were walking through the rubble of the courtyard which a few years before had been the ground upon which the Holy Temple stood. Jackals and other wild animals were wandering to and fro. Rabbi Yehoshua cried bitterly:

> "Look at the desolation that was once the glory of
> Israel!"

Rabbi Akiva began to chuckle. Rabbi Yehoshua turned to him and asked how he could react in such a manner to the tragic consequences of Jewish destruction. Rabbi Akiva explained:

> "In the prophecies of Micha, it is foretold that
> Zion shall be ploughed as a field...[2]
> And Jerusalem shall become a heap of ruins
> And the mountain of the temple
> A mere wooded height."[3]

"But," he continued, "the prophet Zechariah foretold of men and women sitting joyfully in Jerusalem."[4] If Micha's threatening prophecy would not have been fulfilled, I would have doubted Zecharia's prophecy. Now that I have seen with my own eyes the fulfillment of Micha's prophecy, I am certain that Zecharia's prophecy will also be fulfilled in the time of the future redemption."

1 Talmud Bavli Makkot 24 b
2 Micha 3:12
3 Jeremiah 26:18
4 Zecharia 8:4

THE DISPUTATION AT BARCELONA[1]

The Moslems conquered Spain in the early part of the eighth century. The Jews who were living in Spain enjoyed religious and economic freedom under the rule of the Moslem conquerors. For hundreds of years, the Jews participated in Spanish civil society, contributing to the economic, political, scientific and intellectual development of the country. The Jews were businessmen, landowners, court advisors and money lenders. Many Spanish Jews were prominent citizens and boasted of noble lineage and enormous wealth.

After approximately three hundred years, this situation changed. The Moslems began to lose their foothold in Europe. Large areas of the Iberian peninsula were gradually reconquered by Catholic kings. These new conquerors weighed the question of Jewish usefulness to the economy against their allegiance to the Church. They chose to ally themselves with the Church and follow its tenets. One of these tenets was the condemnation of the Jews to eternal wandering and perpetual slavery, making them scapegoats for the crucifixion.

Gradually, restrictions were placed against Jewish artisans and craftsmen, prohibitions were enacted against the Jews bearing arms, mingling with Christians and wearing the same attire. Jews were forced to wear coarse looking clothing decorated with a distinctive badge indicating the wearer was Jewish.

To justify their attitude toward the Jews, the Church devised a scheme whereby its leaders ordered the Jews to publicly debate their beliefs, fully realizing that they were placing the Jews in jeopardy. The Jews were limited in what they were permitted to retort to the judges. These public debates were called 'Disputations'. One apostate Jew, Pablo Christiani, attempted to show that the Rabbis of the Talmud accepted the belief that the Messiah had already made his appearance. He wanted to prove that Judaism was no longer valid. He hoped to gain favor in the eyes of the church fathers, so he issued a challenge to dispute the validity of Judaism. Rabbi Moses ben Nachman, Nachmanides, the RambaN, accepted Pablo Christiani's challenge to debate the issue in order to prove his illogical reasoning.

The RambaN was summoned to the disputation which was held under the auspices of King James of Aragon and church leaders. The King promised the RambaN freedom of speech during the disputation.

The RambaN was born at the end of the twelfth century in Gerona, Catalonia, Spain. He was the descendant of a long line of prominent scholars and he distinguished himself by following in their footsteps. He was one of the acknowledged leaders of Spanish Jewry.

The RambaN began his argument by saying that his life was in the hands of the Church, but his soul was in the hands of G-d. Then he quoted from the prophet Isaiah in his retort to the Church dignitaries who gathered to hear this Jew defend the validity of Judaism by arguing that the Messiah had not yet appeared in the world.

"It shall come to pass in the end of days...
Many people shall go and say
Come, let us go to the mountain of the Lord
To the House of the G-d of Jacob, and He will
teach us His ways and we will walk in His path
For out of Zion shall go forth the law
And the word of the Lord from Jerusalem...
And men will beat their swords into plowshares
And their spears into pruning hooks,
Nation shall not lift sword against nation,
Neither shall they learn war anymore."[2]

The RambaN argued: "Since the advent of Christianity, the world is filled with needless bloodshed and hatred against the Jews. Is this the fulfillment of the prophecy of peace? Is this the vision of the Messianic Age?"

After the Disputation, the Bishop of Gerona asked the RambaN to record his arguments. He agreed and wrote "Sayfer Havikuach" (The Book of Arguments). Several months later, the Bishop brought the RambaN to trial, using the arguments from his book as evidence that the RambaN had blasphemed the church.

During the trial, the RambaN defended himself by saying that he was promised free speech by King James for the duration of the Disputation, and his book reflected these arguments. His defense was rejected. The RambaN realized that his life would be in great danger were he to stay in Spain, so he fled to Eretz Yisrael where he lived the remainder of his life.

There he wrote his comprehensive commentary on the Torah. It is a fundamental source of Jewish philosophical and mystical thought. Many people followed his example and joined him in Eretz Yisrael. The migration from the exile to the Holy Land renewed Jewish settlement in Eretz Yisrael which became a haven for those people who fled from religious persecution.

1 Yaakov Katz and Moshe Hershko. Yisrael V'Ha-a-mim. Dvir Publishing Co. Tel Aviv. 1961.

2 Isaiah 2:1-4

BE CAREFUL HOW YOU SPEAK[1]

The Chmielnicki massacres in the middle of the seventeenth century decimated the Polish Jewish community, one of the most prominent in Europe. The physical and spiritual state of the remnants of Polish Jewry degenerated into helplessness and hopelessness. Even when the Polish king regained control of his land and was able to subdue the Cossack marauders, Jewish suffering did not cease. Bandits roamed the countryside; and "blood libels"[2], previously unheard of in Poland, became commonplace.

These tragedies fostered a messianic movement. It was led by a charlatan who purported to be the redeemer of Israel. Many Jews sold their property, packed their belongings and made travel arrangements to return to their ancestral homeland. These people were deluded into believing that Shabbatai Tzvi was the Messiah. The Rabbis recognized that the Shabbatean movement, the movement formed by the followers of Shabbatai Tzvi, was a cruel hoax and pleaded with the people not to uproot themselves. However, many people refused to heed their advice, uprooted themselves and set sail for Eretz Yisrael. Eventually, under pressure, Shabbatai Tzvi converted to Islam as did many of his followers.

Most Jews realized that their suffering had not come to an end. They had been looking forward to redemption since the destruction of the Second Holy Temple. They prayed three times a day, in the "Amida" (silent devotion) for the rebuilding of Zion:

> "V'techezenah aynaynu b'shuvecha l'tziyon b'rachamim"
> (May our eyes witness Your return to Zion.)

When they recited the "Birkat Hamazon" (Grace after Meals), they pleaded:

> "U'v'nay Yerushalayim eer hakodesh bimhayrah b'yamaynu"
> (Please rebuild Jerusalem quickly in our days.)

The Jews wondered how long they would be forced to wander from country to country; they wondered when redemption would bring an end to the long and bitter exile. They wondered how long their prayers, yearnings, hopes... would go unanswered? The Jews were weary of suffering. They eagerly awaited the era of peace foretold by the prophets.

During the period following the tragedy of Chmielnicki and the delusions of the Shabbatean movement, the Baal Shem Tov appeared on the scene. Rabbi Yisrael ben Eliezer, the Baal Shem Tov, the founder of the Chassidic movement, raised the hopes of Polish Jews by teaching them to serve G-d with great joy, despite their personal feelings of despair. He infused in his disciples the love for Eretz Yisrael and instilled in them a great yearning for the coming of Messiah, the true redemption.

The Baal Shem endeavored to undertake the hazardous journey from Medziboz to Eretz Yisrael many times, but each time his plans were thwarted. He finally realized that it was not meant for his feet to touch the soil of the Holy Land. However, he encouraged his disciples to go to Eretz Yisrael. Each student that journeyed there received a blessing from the Baal Shem before his departure.

One such student received a warning rather than a blessing before his departure. The Baal Shem simply told him:

"Be careful to whom you speak and what you say."

The student met many people along the way, but he never forgot the Baal Shem's admonition to be careful with whom he spoke and what he said. After many months of travel, he arrived in Eretz Yisrael. The first thing that he did was go up to Jerusalem to pray at the "Kotel" (Western Wall). He prayed at the "Kotel" many times during his stay in Eretz Yisrael, each time completely overwhelmed at the beauty of the living stones that were a remnant of Israel's ancient glory. He also visited other holy places in Eretz Yisrael, never forgetting the Baal Shem's admonition.

Once he met a beggar wandering around the ruins of the Holy Temple. The beggar approached him and began to speak:

"You are from 'chutz la-aretz' (diaspora), aren't you?"

The student answered in the affirmative. The beggar continued:

"Tell me how the Jews are faring in the diaspora, in the world outside of Eretz Yisrael?"

The student unhesitatingly answered:

"Life is very difficult; most of the Jews are very poor, but we are managing."

The beggar continued:

"Are the Jewish people studying much Torah?"

The student again answered unhesitatingly:

"Some Jews are great scholars, many could learn more, but we are managing."

The beggar continued:

"How do the Jewish people live?"

"Most Jews live enclosed in ghettos or are confined to special sections but we are managing."

Thereupon, the beggar queried:

"How much are the Jews yearning for redemption?"

"We are yearning, but we are managing."

Upon hearing that the Jewish people were managing their affairs in exile, the beggar disappeared.

This encounter had no special significance for the student. He remained in Eretz Yisrael a few more uneventful months and then he returned to Medizboz.

The Baal Shem was awaiting his return and sent for him the moment he arrived.

> "Tell me," said the Baal Shem, in a disconcerted voice, "Did anything unusual happen on your trip?"

The student did not understand why the Baal Shem was so perturbed. He said:

> "Nothing unusual happened. Soon after I arrived at the "Kotel", I met a strange man, dressed as a beggar. He asked me a few questions, such as what Jewish life was like in the exile, then he disappeared."
>
> "Please tell me what you told the beggar" asked the Baal Shem.

The student answered:

> "I told him that life was difficult but we were managing; that we were yearning for redemption but we were managing."
>
> "You fool," shouted the Baal Shem, "That beggar was Eliyahu Hanavi on his way to fetch the Messiah. I admonished you to watch what you say. When you told Eliyahu that we were managing, he changed his plans. Had you told him that we couldn't wait for Messiah another minute, he would have hastened the coming of redemption. Since you told him that we were managing, he delivered the message to the Heavenly Court that we were still not ready."

1 This story was told at a class in the House of Love and Prayer in San Francisco by Rabbi Shlomo Carlebach. I participated in that class.

2 Blood libels were allegations that Jews murdered Christians in order to obtain blood for the baking of matzot for the Passover holiday. The blood libels were deliberately perpetrated to incite pogroms against the Jews.

EVEN THOUGH HE TARRIES, I BELIEVE[1]

The Trisker Maggid, one of the forebears of the twentieth century Twersky Chassicic dynasty, yearned so much for the coming of the Messiah, that he harnessed his wagon and rode off with his seven year old son one morning in search of the time when ultimate peace will reign in the world. The following story is about his search for the Messiah.

The Trisker Maggid never spoke before he "davened" (prayed) in the morning; he ate very little and he slept even less. When people asked why he slept so little, he answered:

> "It is difficult to sleep when you are hungry."

When asked why he ate so little, he answered:

> "If you slept as little as I do, you wouldn't have the strength to eat; so much the more, if you were sent to bring the Messiah and you knew the world was not ready, could you ever eat or sleep?"

One morning, at the break of dawn, before the Trisker Maggid "davened" (prayed), he asked his seven year old son to harness his wagon to prepare for a journey. The young boy thought it strange that his father would leave the village before the morning prayers, but the boy respected his father and didn't question his actions. The sun shone gloriously that day; the blades of grass stood tall; the trees pointed their crowns toward the heavens and the white clouds raced toward Jerusalem. The two travelled for a while. All through the journey, the Trisker Maggid was elated with the hope of peace and redemption.

Suddenly the Maggid stopped the wagon and pulled it over to the side of the road. He asked his son to remain in the wagon. He alighted and walked over to a humble hut at the side of the road. A smiling man came out to meet him and the two men chatted. At the end of their conversation, the weather began to change. Stormy winds blew and dark clouds covered the bright sun. The blades of grass shrivelled and the crowns of the trees bent over in sadness.

The Trisker Maggid returned to his wagon, jumped in, glanced around to wave farewell to the man, turned the wagon and started homeward.

The young boy leaned toward his father and asked him to explain the sudden change in the weather from the shining light to the gloomy darkness. The Trisker Maggid answered:

> "My son, I went in search of the Messiah. I want you to know that the man I spoke to was the Messiah. He wanted to know if people were truly ready to live in ultimate peace and when I told him that we were still not ready, he advised me to return home and continue to wait."

1 This story was told to me by Rabbi Shlomo Carlebach.

THE PASSENGER MUST BE THE MESSIAH[1]

Rabbi Eliyahu ben Shlomo, the Vilna Gaon, was the undisputed halachic authority of the eighteenth century. As a youngster, he was considered a child prodigy. After his Bar Mitzva, he was recognized as an outstanding Torah scholar. He was master over complex Talmudic texts, elucidating certain tractates with elaborate commentaries. His primary work was a commentary on the "Shulchan Aruch" (the code of Jewish law) written in his own hand, and included in all printed editions of the text.

The Vilna Gaon understood, as Rabbi Akiva understood, the reasons why certain letters of the Torah were adorned with crownlets when they were written on the parchment scroll. He could explain the different musical notes when the Torah was chanted before a congregation. Just as his predecessor, the Rambam (Maimonides), he was also well versed in the secular sciences, in philosophy, engineering and mathematics.

The title "Gaon" was given to him by the people of the Vilna Jewish community as a sign of the awe that they felt for this man who personified religious leadership. When people asked Rabbi Eliyahu, the Vilna Gaon, how he merited this title, he answered in Yiddish:

> "Vilnor, (if you will it) un du vest oychet veren a 'vilnor' Gaon" (you can also become a Vilna Gaon).

The people of Vilna respected the Gaon's devotion to learning. The neighborhood children knew not to play outside the Gaon's 'House of Learning' lest they distract him. His disciples said he never slept more than two hours in any twenty-four hour period. Despite this, he didn't separate himself from the people.

During one point in his life, the Vilna Gaon decided to taste the bitterness of the exile. He wanted to know how the Jews who were living outside of Vilna, Lithuania were faring. He decided that he would wander around Poland and Germany for five years. The Vilna Gaon scheduled his travelling time so that it would not conflict with the time he set aside to study Torah and Talmud. He disguised himself as a beggar in order not to be recognized.

Once, as he was trudging along a road, a peasant stopped his wagon and invited him to ride with him. The peasant was very tired and in exchange for the ride, he asked the "beggar" to take the horse's reins and guide it for awhile. The peasant climbed into the back of the wagon and went to sleep. The "beggar" sat in the driver's seat and directed the horses towards the town. In spite of his disguise as a beggar, the townspeople recognized the driver as the Vilna Gaon.

Suddenly one Jew began shouting:

> "The Messiah is coming, the Messiah is coming!"

Crowds of people ran from their homes and shops and hastened to the synagogue. As they ran, they asked the Jew who was running around the town how he knew that the Messiah was coming. He shouted:

"If the driver of the wagon is the Vilna Gaon, his passenger must be the Messiah."

1 Rabbi Sholom Klass. Tales of our Gaonim. The Jewish Press. Balshon Printing Co. Brooklyn, N.Y. 1967.

THE MIRACULOUS PAPER[1]

Until the third partition of Poland in the last quarter of the eighteenth century, few Jews lived on Russian soil. After the third partition of Poland, Russia became the master over half of the world's Jewish population that was living in the annexed territory. This Jewish population numbered about twelve per cent of the total population of that area. The Russian government was intolerant of foreigners, particularly Jews, but found it impractical to displace the tremendous Jewish population from the land.

Tzarina Catherine the Great, in an official act of the Russian government, decreed that the former Polish kingdom which Russia annexed be called the Pale of Settlement and prohibited Jews from venturing forth from this territory.

The word "pale" is derived from the Latin "palus" meaning a stake or pole, and it was an area surrounded by poles which restricted or confined people who were under the jurisdiction of another country. The Pale of Settlement was located between the Baltic and Black Seas from north to south, and by the surrounding partitioning countries, Russia, Prussia and Austria on the east and west sides. The Jews were confined to small villages in the Pale of Settlement, the western provinces of the empire, to prevent their spreading to other parts of Russia. They were treated as a hostile population, different from other Russian citizens, unworthy of equal citizenship.

The Jews became weary of the hostile treatment of the Russian government. They were concerned for their children's religious development when the Cantonist Laws conscripting Jewish children into Tzar Nicholas I' army for twenty-five years went into effect. They despaired when Tzar Alexander III proposed a plan for the solution to the Jewish problem; death to one third of the population, expulsion for one third and forced conversion for the remainder. When the government began implementing this plan, pogroms broke out in the small villages. The local police would stand idly by as ruffians ransacked houses and burned marketplaces. Oftentimes, whole villages were destroyed. The Jews had earned a meager livelihood as small shopkeepers in the villages of the Pale of Settlement, but the Russian population accused them of economic exploitation. In May, 1882, Tzar Alexander III issued a series of edicts known as the "May Laws"[2] whereby Jews were forced out of the villages into larger towns within the Pale of Settlement. Thousands of Jews were uprooted from their homes and forced to move to the larger towns where they could not earn a living.

More and more, the Jews of nineteenth century Russia trusted in the salvation promised by Israel's ancient prophets. They knew that as long as they lived under the rule of the Tzars, their suffering would never end. Yearning for redemption increased and stories about the coming of the Messiah became very popular.

The following story was told to the Strettiner Rebbe by an innkeeper. It reflects the yearning of Russian Jews for redemption.

Many years ago, Jews were not permitted to live in or travel to Russia unless they were very wealthy. Only on rare occasions were permits granted to Jewish merchants to travel through Russian cities for business if their purpose guaranteed benefit to the empire. On these exceptional occasions, the Jewish merchants were forced to abandon their Jewish garb and don the dress of successful Russian businessmen.

My father was a very wealthy man, and when I was little, I travelled with him on his business trips all over Russia. When we travelled through the Russian countryside, my father and I wore big yarmulkes and we left our "tzitzit" (ritual fringes) hanging out of our garments because we wanted everyone to see our Jewish uniform. When we went to St. Petersburg, we tucked our ritual fringes in and exchanged our yarmulkes for big fur hats so we would look more like Russians than Jews. We always stayed in the best hotel in St. Petersburg.

We had just checked into St. Petersburg's finest hotel when we heard the strange sound of a "shofar" (ram's horn) blowing like a siren from all four corners of the city. My father turned to another hotel guest and asked what the strange sounds were. He was told that representatives of the ten last tribes were encamped outside the city. The Tzar had granted them permission to camp there and two of its generals were hotel guests. The sound of the shofar was the signal for the emissaries to return to their camp. My father and I accepted the explanation with reservation and we went to our room.

The next morning we awoke early and went down to the dining room to drink tea. While sitting in the dining room, we noticed another strange sight. Two Jews, wrapped in "talis and tefillin" (prayer shawls and phylacteries) were proudly standing in a corner, fervently shaking with great devotion as they recited their morning prayers aloud. We were amazed that these Jews were not afraid to be in St. Petersburg. They had a majestic appearance. Just as we were about to go over to talk to them, we heard the sound of the "shofar" again. The Jews quickly put away their "talis" and "tefillin" and disappeared.

That evening we saw the same two Jews walking through the hotel lobby to their rooms, but all their pride was gone. The next morning, my father and I woke up early and went down to the dining room for tea. There we found the two Jews praying, but this time they were so sad, weeping, tears streaming from their eyes. Their prayers sounded like the lamentations of Tisha B'Av (the day mourning the destruction of the Holy Temple) and the supplications of Yom Kippur (the Day of Atonement) combined. My father could no longer restrain himself. He took off his fancy Russian hat, put on his big yarmulke, pulled out his ritual fringes and walked over to the corner where the two Jews were praying. I followed him.

One of the Jews turned around when he realized I was standing near him. He kissed my ritual fringes and put his hands on my head. My father asked them in Yiddish who they were and what they were doing in Russia, but they did not understand. My father tried Russian, but they still did not understand. Finally my father spoke to them in Hebrew which they understood.

One of the Jews told us the following story:

> "We are the representatives of the ten tribes, the ten lost tribes, the ten holy tribes. We were invited to visit St. Petersburg by the Russian government as foreign emissaries to help the Russian government find a way to deal with the Russian Jews. We have spent a great deal of time wandering around the Pale of Settlement, seeing how the Jews live. We have just returned to St. Petersburg and we were waiting for the Heavenly Court to announce the redemption of the Jewish people so we might lead the ingathering of the exiles to Jerusalem, the holy city.
>
> "Yesterday, when we were praying, we had the feeling that the time had arrived for us to fulfill our mission of leading the exiles back to Eretz Yisrael. However, it was revealed to us by the Heavenly Court that the Messianic Age had not yet arrived."

My father said to the two Jews:

> "Please give me a sign that you are representatives of the ten lost tribes."

One Jew took a piece of paper from his pocket and handed it to my father and said:

> "Here is a piece of paper which looks blank. You will not be able to see anything which is written on it until after Yom Kippur. At that time, take the piece of paper outdoors and hold it up to the moonlight. You will be able to see your name written on the paper. If you have very clear eyes, you will be able to see your father's name. If you have clearer eyes, you will be able to see the names of your children, your grandchildren and your greatgrandchildren. If you keep looking, and if your vision is perfect, you will be able to see the day on which the Messiah will come."

My father thanked him for the blank paper and put it in his pocket.

My father was a follower of Rebbe Mordechai Neshchizer. Chassidim say that only three people knew when the Messiah was coming and these three people were our patriarch Jacob, the prophet Ezekiel and Rebbe Mordechai Neschizer. My father gave the paper to his Rebbe, and he told him the story of how he obtained it from the Jews who claimed they were from the ten lost tribes.

We waited anxiously for the sounding of the shofar, signifying the end of Yom Kippur. Immediately we went outside. The Rebbe held a candle in his hand and raised it high. He took the blank paper and held it up to the light of the candle and he began to read what he saw. First, he read his own name, Mordechai ben Dov Baer, then he read the names of his ancestors all the way back to our Patriarch Abraham. Then the Rebbe began to read the names of his children and grandchildren, Yitzchak ben Mordechai, Yosef ben Yitzchak... Then he stopped. He said, "hold the candle very still". He held the paper very close to the candle and he looked once more at the paper. What he saw frightened him and with a trembling hand he put the paper to the flame of the candle and burned it.

To this day, nobody knows what he read on the paper or why he destroyed it, but rumors have it that he saw the date of the coming of the Messiah.

1 <u>Shivchay Tzaddikim.</u> Bayt Hillel Publishing House. Reprinted in New York, 1980. This book is an anthology about Chassidic Rebbes.

2 The May Laws were enacted in May, 1882, under the reign of Tzar Alexander III. The Jews were forbidden to settle outside the towns and townlets. Deeds of sale and leases of real estate in the name of Jews outside the town were cancelled. Jews were prohibited from trading on Sundays and Christian holidays. Police were charged with implementation of these laws which became a source of constant police extortion and harassment of Jews still living in the villages.

A NEW LIGHT WILL SHINE ON ZION[1]

Russian Jewish life continued to deteriorate when a disciple of Rebbe Issachar Ber of Radishitz described the Rebbe's vision of the days preceding the advent of the Messiah.

One Friday night, in the "bayt hamidrash" (study hall) of Radishitz, Rebbe Issachar Ber called Chaim Mayer'l, his best student, to his side. He put his arm around Chaim Mayer'l and walked with him to the window. Both the Rebbe and his disciple peered out the window into the darkness.

The Rebbe asked Chaim Mayer'l:

> "What do you see in the distance?"

Chaim Mayer'l did not reply. Silently, both stood by the window for a long time. The Rebbe began to tremble and then he started to cry. He clutched Chaim Mayer'l tightly. The Rebbe's grasp caused Chaim Mayer'l to shudder. He sensed that the Rebbe's tears were symbolic of the oceans of pain which the Jewish people would suffer from that day until the arrival of the Messiah.

Later on, Chaim Mayer'l's friends asked him what the Rebbe showed him.
He said:

> "the Rebbe showed me "or chadash al Tziyon taeer" (a new light will shine on Zion). He showed me what will happen in the time of the Messiah."
>
> "Then why was he trembling?" asked Chaim Mayer'l's friends.
>
> "He was trembling," replied Chaim Mayer's, "because he also showed me all the pain and suffering until that great day arrives."

1 This story was told by Rabbi Shlomo Carlebach at a concert in West Palm Beach, Florida, in March, 1981.

EVEN IN DESPAIR, I BELIEVE[1]

During the many years of historical research on this book, I dreamed of being able to visit the places in Europe which were the "scene of the action". In July, 1982, I had the opportunity to fulfill this dream. My visa application to the Czechoslovakian government stated that I was a teacher involved in historical research, so my request was granted.

With keen anticipation, I boarded a flight at Kennedy International Airport, destination eastern Europe. As the pilot guided the plane across the Atlantic Ocean, I tried to imagine what I would discover in the "old country". I knew that Jewish settlement in eastern Europe, particularly Prague, Czechoslovakia, began more than a millenium ago.

Prague and the areas comprising Bohemia-Moravia, in central Europe were great centers of Jewish settlement for the past thousand years. Its leaders were considered world famous scholars and teachers: Rabbi Yehuda ben Bezalel Low (the Maharal of Prague), Rabbi Yeshayahu HaLayvi Horowitz, Rabbi David Oppenheim, Rabbi Yechezkel Landau (the Nodeh B'Yehuda).

When the brutalities from the German occupation ended, many Jews, their concentration camp numbers branded into their arms, returned to Prague to pick up the pieces of their shattered lives. Many of them had been interred in Theresienstadt, a concentration camp ten miles from Prague. Little did they know that the religious freedom for which they so desperately longed would again be curtailed by the Communist regime which emerged as the most dominant political party in the new coalition government of Czechoslovakia, a landlocked country in central Europe, surrounded by Poland, East Germany, Russia and Hungary.

By the 1950's, the economy was nationalized, political and cultural liberties were curtailed and purge trials became common. Acceding to public protest, some restrictions were eased in the intervening years between 1950 and 1968 when Russia invaded Czechoslovakia and created the Czechs Socialist Republic.

The following stories are about some of the people I met in the "alte-neu" shul and the adjacent Jewish community center in the city of Prague. They are among the 6,000 survivors of a dying community, victims of a government bent upon spiritually destroying the remnants and their children. They are longing for Messiah, for the redemption of the Jewish people from all the lands of modern bondage.

> "And it will be on that day,
> A great shofar will sound,
> And all the lost souls from Assyria and Egypt[2]
> Will come to bow before the Lord on His Holy Mount, in Jerusalem."[3]

These people live on or near the site of the old Jewish ghetto.

It was very easy to locate this area because many stories of the Prague ghetto mention the Ultava River which flows in a semi-circular path around the ghetto. Today, this area is called the State Jewish Museum in Prague,[4] a section of winding streets of about three city blocks by six city blocks.

A young man entered the kosher dining room of the Jewish community center as I was eating my lunch that Thursday afternoon. He was wearing a menorah around his neck. I knew that if I wanted to find out any information, I would have to ask questions. I took a chance and tried to strike up a conversation, not knowing in which language to communicate. To my surprise, the young man spoke English beautifully. He explained that he had a doctorate in English from the University of Prague, but he earned a livelihood washing storefront windows whenever such work was available to him. He told me that he had two strikes against him: being Jewish and being an active member of the human rights movement which began in Czechoslovakia after the Helsinki Agreement in 1977.

He is married and has two young children that he wants to raise Jewishly, but the community offers no educational resources, except for Sabbath services, a Passover community Seder and High Holiday services. He is learning about Judaism from books that miraculously escape censorship. He believes that most of the books, sent to him by friends he has made who have visited Prague, are confiscated by the authorities. He explained that he chose to be identified with the small Jewish community when he was circumcised at the age of twenty-two. Although he was born a Jew, he was not educated in the tenets of Judaism; his father rejected the Jewish religion after his experiences in Auschwitz.

My Czech friend is waiting for the opportunity to escape from behind the iron curtain. He hopes the opportunity will present itself before he is jailed for his "human rights" activities.

I found that there were two types of Jews living behind the iron curtain. The first type of Jew wants to advance politically, economically or socially and chooses to assimilate to insure his physical security and his job security. The second type of Jew is so imbued with Jewish survival, with "hanging on" until the arrival of the Messiah, that he is constantly asking himself why and how he survived. This type of Jew is identified as a Jew because he believes that the outside world will always be hostile to the Jew in one way or another. While he lives, he is determined to live the best Jewish life possible.

Attending the Sabbath services in the "alte-neu" shul, part of the Jewish State Museum during the week, was an extraordinary experience. The remnants of the House of Israel draw strength from visitors, while the visitors gain inspiration from the few Jews who are "hanging on", brave enough to be identified as Jews by attending Sabbath services. This creates an atmosphere of reverence.

An elderly gentleman, about eighty years old, ascends the pulpit in the center of the shul. He serves the congregation by reading the Torah and chanting the services. I imagined the beautiful melodies he sung to be the same which had been sung in this synagogue for centuries. The people like him. He lives with his wife in an apartment a distance from the "alte-neu" shul, but has a room above the Jewish community center that he uses for the Sabbath and holidays. He wonders how the next generation will survive Jewishly without a leader to teach them the melodies for the prayers and the cantillation for the Torah. His two sons have fled and live in the West. Only his wife is permitted to travel to see their children. For reasons only known to the state government, he is not permitted to leave the country. It is apparent that he will never see his sons again.

There are three prayer services during the twenty-five hours of the Sabbath. I noticed that the elderly people attended all three services, but I was amazed that six or seven young men were also "regulars". I wondered about their commitment, so I sat down next to a young man with a "kipa seruga" (a crocheted head covering, name neatly stitched into the border, Israeli style). He had attended the Friday night and Sabbath morning services, and appeared again when the group gathered for "shalosh seudot" (the third meal of the Sabbath) and the concluding service. The elderly gentlemen were seated around a table, sharing a meal of challot, buckwheat and cherries. The younger men sat in a row of chairs arranged behind their elders. I began to talk to the young man next to me. He spoke to me in Hebrew, entirely self taught. I asked him if he desired to immigrate to Eretz Yisrael. He answered in a very soft voice that it was not in the best interests of the State to let him go. In the background, the people were nibbling at their food. Some were singing haunting melodies as they bid farewell to Queen Sabbath. The melodies lingered as darkness descended over Prague. Suddenly, the young man turned to me and said:

> "Even in this place, where it does not seem that there is a Jewish future, 'ani maamin', I believe with perfect faith in the coming of the Messiah, and though he tarry, I will wait daily for his coming, and never lose faith."[5]

1 This is an original story by the author.

2 In the twentieth century, the iron curtain countries are considered lands of Jewish enslavement.

3 Isaiah 27:13

4 "Jewish State Museum in Prague was founded on the basis of a decree issued by the government of the Czechoslovak Republic in 1950 when, in connection with the beginning of the building up of a developed socialist society, the Czechoslovak Socialist State itself took over the task of caring for museum collections and other cultural values in order that they might be made widely accessible to the public. The collections in the charge of the State Jewish Museum come from the property of individual Czech or Moravian Jewish religious communities and synagogues which were forcefully liquidated by the Nazis during World War II on one hand and from systematic collections activity carried out in the post war period on the other hand."
Guidebook. Introduction to the Jewish State Museum in Prague.

5 Maimonides. Shlosha Asar Ikarim (Thirteen Principles of Faith).

THE ETERNITY OF THE JEWISH PEOPLE

נצחיות עם ישראל

CHAPTER EIGHT

"Am Yisrael chai, Ode Aveenu chai"

(The Jewish people live because Our Father in Heaven lives).[1]

In previous chapters, major historical themes, such as love of Torah, martyrdom, repentance, love of Eretz Yisrael, and yearning for the Messiah have been explored through the media of stories. This last chapter deals with the mystery of Jewish survival or the eternity of Am Yisrael.

Nations have marched across the frontier of western civilization and disappeared beyond the horizon. They ascended upon center stage, performed their role and descended, fading into time. Contrary to the laws of history, the Jewish people have not vanished as have other world conquerors. One of the reasons for Jewish survival is the Eternal mission for which the Jewish were chosen.

The Jewish people made a covenant with the Eternal G-d three thousand five hundred years ago when they stood at the foot of Mt. Sinai to receive the Torah. The oath of the covenant is recorded in the Torah:

> "And Moses took the book of the covenant and read in the ears of the people and they said, 'all that the Eternal has spoken we will do and we will listen.'"[2]

The Jewish people swore by their children and their children's children that they would be guarantors of the Torah for all future generations.[3] Since that time, Jews have attempted to conduct their lives in a way prescribed through the "mitzvot" in the Torah because the "mitzvot" are simply G-d's directions on how a Jew is supposed to act.

The Jewish people were "chosen" for the prophetic mission of teaching G-d's moral law to the world.

> "I will give thee for a light unto the nations,
> That my salvation may be unto the end of the earth."[4]

We, the descendants of those who stood at Sinai, bear the responsibility for the continued development and perpetuation of this prophetic mission. As long as we are loyal to this prophetic mission, we are able to survive from generation to generation, century to century, millenium to millenium: an Eternal people representing an Eternal G-d, faithful to an Eternal Law.

> "For the Eternal will not forsake His people for His great Name's sake, because it has pleased the Eternal to choose you as his people."[5]

Faithfulness to Torah, the Eternal Law, is the Jewish way of organized resistance to those who would destroy us. It is an undeniable fact that Torah provides the spiritual weapon for Jewish survival. Shimon Hatzadik said:

> "The Jewish world is upheld by three pillars, the pillar of Torah study, the pillar of prayer, and the pillar of good deeds."[6]

Torah has enabled the Jewish people to cope with intermittent outside destructive forces that nibbled away at the branches of the tree representing Jewish life. True, there were occasions in the Jewish past when most of the branches came dangerously close to annihilation, yet the roots remained strong enough to survive the most horrible onslaught.

A lovely metaphor compares the Jewish people to sapphires and the nations of the world to anvil and hammer.

> "For when placed on the anvil and struck with the hammer, the sapphire remains intact, while the anvil is split and the hammer is broken."[7]

Thus also is the fate of the nations that attempt to crush Israel between anvil and hammer. The nations are broken and Israel lives on.

Today, the invincible Jewish spirit is being threatened by an inner destructive force that is nibbling away at the roots of Jewish life, robbing the tree of its strength.

This threat is ignorance of Torah, and it is responsible for the high statistics of intermarriage and assimilation. Jewish ignorance erases the difference between Jew and non-Jew. In western society, where political equality before the law is guaranteed, Jews must maintain a Torah education to justify their "chosenness" in order to survive.

Always, when Jews feel frail and feeble, a new source of strength emanates from hidden places. In recent decades, notably the sixties and seventies of the twentieth century, a "teshuva movement" (returning to Jewish roots) has developed. This rejuvenation of Jewish roots is amost unparalleled in the annals of Jewish history. It is being led by the young, "and a young child shall lead them"[8] who are searching for their roots, for meaning and truth in their lives. They are finding meaning through a commitment to Eternal Torah.

Jewish survival is rooted in Torah, in the glory of the past and in hope for the future redemption. We must never despair. We must never stop fulfilling our prophetic mission. We must: "Only do justice, love mercy, and walk humbly with our G-d."[9] We must refuse to die.

> "I shall not die, but live and declare the wondrous deeds of the Eternal."[10]

The stories in this chapter depict various ways in which Jews survived. We must go forth with faith in the future, knowing that there will always be a tomorrow.

1 Genesis 45:3 (Paraphrased)
2 Exodus 24:7
3 Shir Hashirim Raba
4 Isaiah 49:6
5 I Samuel 12:22
6 Avot 1:2
7 Trude Weiss Rosmarin. Jewish Survival. Philosophical Library. New York. 1949.
8 Isaiah 11:6
9 Micha 6:8
10 Psalms 118:17

YOU MAY BURN OUR BOOKS, BUT WE WILL SURVIVE[1]

Since the destruction of the Second Holy Temple in 70 C.E., the Jewish people were painfully aware that they would have to survive a long and arduous exile. About one hundred years after the destruction, Rav, the great Babylonian Amora (the scholars who compiled the Gemara) realizing the conditions under which Jews were living, said:

> "Educating children in the ways of Torah takes precedence over rebuilding the Holy Temple."[2]

He was convinced that the Jewish people could survive without the Holy Temple if they were in possession of the Torah, their treasured heritage.

The Jewish people have been called the "People of the Book." Throughout the centuries, the concept of the "Book" has expanded to include not only the Torah, but the Talmud and Rabbinic literature. The Talmud is central to Jewish thought, but the non-Jewish world believed that certain passages in it were anti-Christian. The more the Jews involved themselves in the development of Talmudic thought, the more the non-Jewish world showed their resentment against the Talmud.

Since the Jewish people were foreigners in hostile lands, they were at the mercy of the rulers and clergy. Throughout the Middle Ages, it was commonplace to publicly burn the books of the Talmud. In 1263, the Talmud was publicly burned in Barcelona, Spain. Pope John XXII ordered the public burning of the Talmud in Rome in 1322. In 1509, the Talmud was publicly burned in Germany. In 1609, the Jewish library at Cremona, Italy, was plundered and 12,000 holy books were burned. The intent of the Christian clergy was to annihilate the Jewish people and their holy books, but they did not succeed in their evil intent. The Jews responded by printing more and more holy books.

The Jews had settled in Poland during the thirteenth and fourteenth centuries, migrants from religious persecution in western Europe. They were granted equal citizenship, and religious life flourished for hundreds of years. However, there were a few exceptions to this overall feeling of security and contentment.

This is the story of one Polish king who firmly believed that the books of the Talmud were dangerous and should be publicly burned. Accordingly, he issued a decree to carry out his convictions. Miraculously, the evil decree was thwarted and the Jews of Cracow were able to save their books of the Talmud from being consumed.

Cracow was a major Polish city with a large Jewish population in the early part of the seventeenth century. Its chief Rabbi was Rabbi Natan Nota Shapiro, author of Megalleh Amukot (Revelation of Hidden Secrets) a book of numerological exposition on the letters of the words of the Torah.

When Rabbi Shapiro realized that his days were numbered, he put a plan into action to assure that his successor as chief Rabbi of Cracow would be worthy of that position. He called together his trusted disciples and advised them that he had chosen "Moshe the water carrier" to succeed him. He instructed them to search the city for Moshe and inform him of his decision. Rabbi Shapiro's disciples were astonished because they only knew one "Moshe the water carrier" and he was reputed to be the town ignoramus. Nevertheless, they followed the Rabbi's wishes and went to search for Moshe. When they found him, they told him the wishes of Rabbi Shapiro. Moshe refused to accept the position. He said that he did not know the duties of a Rabbi, nor did he want to leave his customers. The disciples persisted. They convinced Moshe to accept the position as Rabbi Shapiro's successor by saying that surely Rabbi Shapiro knew what he was doing. When Rabbi Shapiro passed away, Moshe assumed the position of Chief Rabbi of Cracow.

The first problem that Moshe had to face was the decree of the Polish king to gather all the books of the Talmud in Cracow in a month's time and burn them publicly in the city square. Notices of the royal decree were posted on the buildings and trees of Cracow. The Jewish people had been living in Poland in relative peace and security. They had not been subject to such a discriminating edict for a long time. Unprepared to deal with this dangerous situation, they ran to consult their new Rabbi. Rabbi Moshe was dismayed at this sudden change of events. He did not know how to advise his people, so he said:

> "Give me time to think this over, and at the end of a month I will decide upon a course of action."

As soon as the Jews left his house, Rabbi Moshe disappeared.

The townspeople searched for him, but they could not find him. They searched the city and checked all the inns in the neighboring villages. When they could not find their new Rabbi, they decided to wait the month. At the end of the month, the innkeeper of a neighboring village let it be known that a stranger was staying at his inn. The Jews of Cracow hastened to send a committee to the inn to find out who the stranger was. To the committee's amazement, they found Rabbi Moshe in a secluded corner room of the inn. He was sitting at a table, heaped high with the books of the Talmud. He apologized to the committee for his disappearance and said:

> "When the edict was issued, I did not know how to deal with it, so I secluded myself in this inn to find a solution to the problem. During this time, I reviewed the entire thirty-seven tractates of the Talmud with Rashi's commentary and I prayed to the Almighty that through the merit of my learning Talmud, a solution would be forthcoming. It pleases me to advise you that it has been revealed to me through a dream that the king who supported scholarship has had second thoughts about destroying scholarly works and has decided to rescind his decree. Now we can return to Cracow assured that our books of the Talmud will not be burned."

The committee gazed in amazement at their Rabbi, the modest water carrier. As they approached the city limits of Cracow, they noticed that the royal decrees were being removed from the trees and buildings upon which they had been posted. The Jewish community of Cracow then realized that Rabbi Shapiro's choice of a successor was based on the knowledge that the humble water carrier was a great scholar. In his merit, the books of the Talmud belonging to the Jews of Cracow were saved from being consumed.

As long as "the Book" is not destroyed, the Jewish people will remain indestructible.

1 Rabbi Yitzchak Komarna. Haychal Haberacha. This book is a commentary on the Torah. Stories of the past are interwoven between the commentaries. This story was told by the Rebbe about 130 years ago.

2 Talmud Bavli Megilla 16 b

KEEP MY MEMORY ALIVE[1]

Jewish settlement in Prague is recorded as early as the beginning of the tenth century. As the years passed, there were intermittent episodes of religious persecution alternating with religious toleration. Martyrdom during the Crusades, blood libels, exorbitant taxes, burning the ghetto, and periodic expulsions are part of the history of the Prague ghetto. However, depending on the whims of the rulers of Prague, expulsion edicts were rescinded occasionally, Jewish liasons were appointed between the Jewish community and the king, and the walls of the ghetto were expanded to accommodate an increasing population.

By the time Rabbi Yehuda Lev ben Bezalel, the Maharal, became the chief Rabbi of Prague at the end of the sixteenth century, there were approximately four thousand Jews living in the Prague ghetto.

This story is about a Jewish soldier who beseeched the Maharal to institute the annual recitation of Kaddish for him after Kol Nidray on the eve of Yom Kippur.

Kaddish is the doxology, the prayer praising G-d, which is recited either after the conclusion of a chapter of Torah, Psalms or Talmud, or by a mourner who declares his faith after the dire pain of losing a loved one. The person who recites the Kaddish after concluding a chapter of Torah, Psalms or Talmud affirms his faith in G-d and in the eternity of the Jewish people. The mourner who recites the Kaddish reaches out to G-d for strength and consolation, knowing that the memory of his loved one links the span of time.

The soldier in this story wanted to be certain that his memory would be venerated by future generations. He wanted the Kaddish recited after him to be a spiritual bridge linking the past with the yet unborn generations of the future.

The story begins with an event that occurred late one Friday afternoon, two days before Yom Kippur.

A wealthy man named Mordechai, who lived in the Prague ghetto, was awaiting the birth of his first child due in a few weeks. He did not anticipate an early birth and therefore made no preparations. Mordechai became alarmed when his wife went into heavy labor shortly after the Sabbath began. He ran to the midwife for help. She told him to prepare hot water for the delivery of the child.. Since he could not light a fire on the Sabbath to heat the water, which was essential for the delivery of the child, he ran to the gates of the ghetto, which was located on the banks of the Ultava River, in search of a non-Jew who would light a fire for him.

Mordechai found the lieutenant, who was stationed at the entrance to the ghetto, walking along the river's edge. Mordechai pleaded with the lieutenant:

> "Today is the Sabbath and I am not permitted to
> light a fire. My wife is in labor and the midwife

> is in desperate need of warm water. Would you please send me a non-Jew to light a fire in my house. I promise to pay you on Sunday."

To Mordechai's surprise, the lieutenant responded in a warm, gentle and friendly way.

> "I know who you are," he said, "return to your wife and attend to her. You can depend on me to help as long as I am stationed here."

The lieutenant kept his promise to Mordechai and sent a soldier to Mordechai's house to light the fire to heat the water. After the baby was born, the midwife went home to take a rest for awhile. She planned to return later to check on the mother and the baby. A few hours later, the midwife began walking back to Mordechai's house. On the way, she saw the soldier who had lit the fire laying dead in the street. Frightened and excited, she entered Mordechai's house and hysterically related what she had seen.

Mordechai realized that finding the dead soldier in the ghetto was a serious matter that would certainly be blamed on the Jews. Not knowing what to do, he ran to the "Parnas" (the communal leader of the ghetto) to tell him what had happened. The frightened "Parnas" consulted with the Rabbi of the ghetto, the Maharal of Prague. The Maharal advised the "Parnas" to tell Mordechai to seek out the lieutenant and explain what had happened.

> "The lieutenant will tell Mordechai what to do," said the Maharal calmly.

It was four a.m. on the Sabbath day and a trembling Mordechai knocked on the door of the lieutenant's house, woke him and told him the story. The lieutenant advised Mordechai to place a bottle of whiskey in the hand of the dead soldier, carry the corpse to the edge of the ghetto and push it under the ghetto gates. He suggested to Mordechai that when the corpse would be found it would appear as if the soldier had drank himself to death.

Mordechai followed the lieutenant's advice and then he went to tell the Maharal what he did.

The Maharal spoke very firmly to Mordechai.

> "You know that it is permissible to desecrate the Sabbath if a human life is in danger. The life of your wife and your new baby were in jeopardy. You were permitted to light the fire to heat the water for the delivery of the baby. Had you lit the fire, the soldier would not have been involved. You have endangered the entire Jewish community by your action. This is the last Yom Kippur that you may spend in Prague. You are excommunicated immediately following Yom Kippur."

Mordechai sadly left the Maharal.

The Maharal, fearing that the incident was not yet over, went to the "alt-neu-shul". As he arrived, the lieutenant who had been waiting for him followed him inside the synagogue and begged to speak to him privately.

"You already know about the dead soldier. Now I need a favor from you. I want you to promise me that you will recite Kaddish after me each Yom Kippur. I need a few minutes of your time to explain to you the reason for this strange request."

The Maharal listened intently as the lieutenant told this story.

"I am now thirty years old. I was born in Poland to a scholarly, rich and noble Jewish family. As was the custom when I reached the age of fourteen I was married - to a beautiful girl. By the time I was fifteen, I was a wealthy man and the father of a child. Due to unforseen circumstances, my wife and I only had five years of happinesss together.

"My father had two brothers living in Amsterdam. Ten years ago, he received a letter from one of his brothers that the other had died and his will instructed that his vast estate be left to my father. My father asked me to go to Amsterdam to settle his late brother's estate.

"When I arrived in Amsterdam, I could find lodging only on the top floor of an inn. The first night I was in Amsterdam, a fire broke out in the inn. I prayed to G-d to save my life and jumped to the ground, leaving all my identification papers in the burning inn. I was knocked unconscious by the jump and when I awoke I found that Moshe Tauber, a local investment counsellor and a friend of my deceased uncle, had taken me into his house.

"I trusted him and told him my story. He promised to arrange for an attorney to send to Poland for new documents to prove my identity. I never received the documents, so I assumed that my family was unable to reach me. In the meantime, my uncle's estate was probated and in order for me to receive the money, Moshe Tauber put up a guaranteed bond, swearing to my identity. I received the money that was left to my father. Throughout the long time it took to probate the will, I remained in Moshe Tauber's house. I had no reason to mistrust him, so I asked him to invest the money. Moshe Tauber took the money from me, but to my great disillusionment, he disappeared the next morning.

"Having lost my 'trusted' host and my money, I decided to return to Poland. It was during the hostilities preceding the Thirty Years' War, and since there was fighting in the area in which I was travelling, it took four years to arrive in my home city and search out my old house.

"I arrived at night and located my old house, hoping that my wife and baby were alive and safe. The door to the house was unlocked, so I entered and walked up the stairs directly to the bedroom that my wife and I had shared. My pulse raced with deep anticipation at the prospect of seeing her again. As I bent to kiss the forehead of the lady in the bed, I was grabbed by two strong men. I had walked into a trap set by a Christian nobleman who suspected his wife of infidelity. I shouted my innocence and the noble lady screamed that she would not be unfaithful to her husband with a lowly Jew, but to no avail. I pleaded with the nobleman to permit me to prove I was not to blame for his wife's infidelity, but the nobleman sent for the local parish priest. When the priest arrived, they took me to a secluded room in the large house and questioned me for hours. The priest told me that if I converted, he would turn me free; otherwise, he would make certain that I would be tried in a court and beheaded. He also told me that my wife and baby were dead. I did not believe him.

"Burning with desire to find my wife and baby, I reluctantly consented to baptism. After the baptismal ceremony, the nobleman and the priest arranged for me to ride, bound like a hostage, on a wagon headed for Prague. They instructed the wagon driver to abandon me at the gates of the city. I arrived in Prague with no way of earning a livelihood. Since the intermittent hostilities which preceded the Thirty Years' War were raging, it was easy for me to enlist as a soldier. Yesterday I was promoted to the post of lieutenant for the Prague ghetto. I do not know if I will ever find my wife or baby or if there will be anyone to say Kaddish after I die. Whatever happens, please promise that you will always say Kaddish after me."

The Maharal not only promised to say Kaddish himself, but assured the lieutenant that the recitation of Kaddish would become part of the "Kol Nidray" prayers each Yom Kippur. The Maharal also added his blessing:

"I bless you that you rise to a position of authority so that you can save many Jewish lives!"

On Yom Kippur fifty years later, the grandson of the Maharal of Prague noticed an old general walk into the "alt-neu shul". The general remained standing throughout the service. After the "Kol Nidray", the Rabbi rose to recite the Kaddish. Suddenly there was a heartbreaking cry from the rear of the synagogue.

When the prayers were concluded, the old general approached the Rabbi and asked to speak to him for a moment. He said:

> "No doubt you are aware of the story behind the custom of saying Kaddish after "Kol Nidray" from your grandfather. I assume that you must realize that I am the lieutenant and I want to thank you for saying Kaddish. Your grandfather's blessing was fulfilled. I saved many Jewish lives throughout my career as a soldier. I want you to know that I recently found the burial place of my wife and baby. Now I have a premonition that this is my last Yom Kippur in this world. I am going to the village where they are buried so that when I die I might be buried near them. Please do not forget to continue saying Kaddish after me."

The custom of reciting Kaddish after the "Kol Nidray" in the "alt-neu shul" remains until this day.

1 Op. Cit. p. 37. Sippurim Prager Sammlung: Judisher Legenden.

WHERE THERE IS LIFE, THERE IS HOPE[1]

Inherent to Judaism is the faith that Jewish life would never cease to exist. During the middle of the seventeenth century, Jewish life was imperiled by the Chmielnicki pogroms and the hordes that followed Shabbatai Tzvi, the false Messiah. The candle that had burned steadfastly through the years of exile began to flicker.

Then, an infusion of hope became visible upon the horizon. Rebbe Yisrael ben Eliezer, the Baal Shem Tov, founder of the Chassidic movement, appeared on the Jewish scene and turned despair into joy. The Baal Shem Tov believed that the highest service of G-d was through happiness, even in moments of despair.

He laid the foundations for the spread of the Chassidic movement through eastern Europe. Upon his death, the mantle of leadership was passed to his disciple, Rebbe Dov Baer, the Maggid of Mezeritch. Many Chassidic communities were started by disciples of these charismatic leaders. Each community had its own particular style but each shared the common goal of revitalizing Jewish life.

Chassidim were drawn to Ger, Bobov, Berditchev, Munkatch and a myriad of other Chassidic communities. Some Chassidim saved money all year to be able to spend one Sabbath or a holiday at their Rebbe's table, singing melodies and listening to his Torah discourses. The songs of the Rebbes were reminiscent of the singing of the Levites in the Holy Temple. The Torah and Talmudic discourses were oftentimes illustrated with parables. The Rebbes told stories to inspire noble action. Visiting their Rebbe was an uplifting experience for the Chassid.

The setting of the following story is the Chassidic community of Medizboz. There, Rebbe Boruch carried on the tradition of his grandfather, the Baal Shem Tov. His home was always open to guests. He particularly invited guests to celebrate the Sabbath or holidays with him. His guest on this particular Sabbath was Rebbe Layvi Yitzchak of Berditchev.

Rebbe Layvi Yitzchak of Berditchev always refused Rebbe Boruch's invitation to join him for the Sabbath. Rebbe Boruch never gave up trying to convince Rebbe Layvi Yitzchak to spend the Sabbath with him. One Thursday, Rebbe Boruch received a rather surprising message from Rebbe Layvi Yitzchak. The message informed Rebbe Boruch that Rebbe Layvi Yitzchak would come to Medizboz for the following Sabbath.

Rebbe Layvi Yitzchak arrived in Medizboz Friday afternoon. Rebbe Boruch was overjoyed to greet his honored guest and welcomed him warmly. Then Rebbe Boruch said to Rebbe Layvi Yitzchak:

> "Please don't be surprised with anything that happens here over the Sabbath."

The two great Rebbes then proceeded to prepare themselves to welcome the Sabbath Queen.

Both Rebbes wore white "kittels" (robes) and wrapped themselves in a special Sabbath "talit". Each "talit" swayed gently in the evening breeze as the Rebbes walked toward the synagogue to welcome the Sabbath. Both Rebbes prayed with great fervor as they anticipated the comradeship of sharing the joy of the Sabbath together.

After they finished praying, the two Rebbes walked slowly, side by side, to the home of Rebbe Boruch. As soon as Rebbe Boruch entered his house, he began humming a soft melody. Gradually, Rebbe Layvi Yitzchak and Rebbe Boruch's disciples, who were also his guests, joined in. The melody grew in intensity until the walls of the house echoed with the reverberation of Rebbe Boruch's Sabbath song. Slowly Rebbe Boruch's feet lifted him off the ground, and as his body moved to the rhythm of the music, the space between his feet and the ground widened. Rebbe Boruch was leaping so high, he was practically flying in honor of the Sabbath. He stopped for only a few moments to make Kiddush, and then he continued with his singing and dancing, his disciples joining him in every beat. He paused again to wash his hands in order to share the Sabbath challot (twisted loaves of bread) with his guest and his disciples. Rebbe Layvi Yitzchak stood by and watched in awe as Rebbe Boruch danced before each course of the Sabbath meal, resting only long enough to eat. The fervor of the dancing grew more and more intense. As the soup was served, one disciple bumped into the dancing Rebbe and spilled soup on his Sabbath "talit", but no words were said and the joy continued.

After the Sabbath there was tremendous commotion as the disciples vied for the privilege of washing the soup stains from Rebbe Boruch's "talit", but Rebbe Boruch insisted that the soup stains remain. When Rebbe Boruch passed away, his special Sabbath "talit" was passed down to his son who passed it down to his son, who passed it down to his son. For four generations, Rebbe Boruch's soup stained Sabbath "talit" was passed down. At the end of the fourth generation, the Holocaust occurred.

Rebbe Boruch could not foresee the destruction of European Jewry in the Holocaust, and his great great grandson, the Rebbe of Munkatch was consumed in the flames. Had the Rebbe of Munkatch been able to visualize the future beyond the Holocaust, he would have conceived of a new generation of Jewish men standing at the "Kotel" in fervent prayer, wrapped in "Talitot" (the prayer shawl symbolizing Jewish royalty), praying for the restored glory of ancient Israel on the soil of her ancestral homeland. However, the Rebbe of Munkatch could not visualize the future; so he ordered his disciples to bury him with the holy soup stained Sabbath "talit" which belonged to his great grandfather. He thought that the Holocaust would mark the annihilation of the Jewish people.

If only the Munkatcher Rebbe could have lived to see the rebirth of the State of Israel, he would not have buried that "talit". Maybe one of his disciples would be wearing that soup stained "talit" at the "Kotel" today.

1 Rebbe Chaim Eleazar Shapira, the Rebbe of Munkatch. Sayder Shana Ha-acharonah. Reprinted by the grandson of the Rebbe of Munkatch, Rebbe Tzvi Natan David Rabinovich, to honor the memory of his grandfather, in New York. 1968.

THEY DON'T BEAT JEWS ALL NIGHT[1]

The setting of the following story takes place in the Chassidic community of Bobov. Rebbe Sholom Bobover learned that the soul of one of his loyal followers was filled with anguish. He set out to kindle hope in Shmuel's heart by telling him that "they don't beat Jews all night".

Rabbi Sholom Bobover had a very tall, strong "shammas" whose name was Moshe David. One day the Rebbe instructed Moshe David to run to Shmuel's house and tell him that it was urgent the Rebbe see him immediately.

Shmuel, a loyal disciple of Rebbe Sholom Bobover, was a very wealthy man. He lived in a luxurious mansion and enjoyed a very high standard of living.

It happened that Shmuel's business began to fail and he was on the verge of bankruptcy. The police threatened to evict him and his family from the mansion. He could not share his despair with either his wife or the Rebbe, so he decided the best way out of the dilemma was to commit suicide. He bought some poison, locked himself in his room and ordered that he not be disturbed under any circumstances for three days.

It was at that particular moment that Moshe David arrived at the mansion. The servants refused to admit him to the house, so he forcibly pushed his way through the front door, ran up the steps to Shmuel's room and broke down the door. He grabbed Shmuel in his arms and carried him to Rebbe Sholom Bobover's study hall. Moshe David lowered Shmuel into the chair next to where the Rebbe was learning. Rebbe Sholom Bobover turned to Shmuel and calmly said:

"I want you to listen to this story.

> "Once when Rebbe Zusia and Rebbe Elimelech were travelling, they had to stay overnight in a Polish inn. Since the inn was very crowded and an extra bedroom unavailable, the innkeeper made the two travellers comfortable by setting up a bed for them in the corridor. When they decided to go to sleep, Rebbe Zusia slept on the outside.
>
> "That night, there was a party in the inn, and when the guests became very drunk, they noticed two Jews sleeping in the open corridor. They decided that it would be great fun to beat them up. Two drunks went upstairs to the open corridor and carried Rebbe Zusia down to the dining room where they proceeded with their game of Jew beating. When they satisfied their lust, they brought a bloody Rebbe Zusia back to the bed.

"Rebbe Elimelech thought that the hooligans might try their game of Jew beating again, so he moved over to the outside to Rebbe Zusia's place and let Rebbe Zusia rest in his place on the inside. Rebbe Elimelech was not surprised when the two drunks came up the stairs again. He figured they would take the man on the outside, but he miscalculated. The drunks grabbed Rebbe Zusia roughly and carried him downstairs for a second beating.

"Toward daybreak, the drunks brought Rebbe Zusia back upstairs and threw him on the bed. His eyes were bloodshot; his face was cut. His muscles ached from the horrible beating.

"Rebbe Zusia turned to Rebbe Elimelech and said:

> 'I want you to know that you can't escape the beatings, but they don't beat Jews all night.'"

Rebbe Sholom Bobover turned to Shmuel and said:

> "Now that I told you this story, go home!"

Shmuel returned to his luxurious mansion, trying to understand the significance, not only of the story, but of his being forcibly carried to the Rebbe.

The next morning, an unexpected messenger arrived at Shmuel's luxurious mansion. He carried a great deal of money, repayment of a ten year old debt which Shmuel had long written off as a loss. Shmuel realized that he could put his affairs in order with this money, but before he did so, he ran to Rebbe Sholom Bobover and said:

> "Rebbe, now I understand the meaning of your story that 'they don't beat Jews all night'. As long as there is life, there is hope that tomorrow will be better."

1 This story was repeated by word of mouth from one Bobover Rebbe to another.

DO NOT DESPAIR[1]

Kishinev was a city in Bessarabia, Russia, located in the southern section of the Pale of Settlement, bordering the Black Sea on the south and Austria-Hungary on the west. By the nineteenth century, the Jews of Kishinev numbered one third of the general population of the city, around 50,000. The Jews earned their livelihood as craftsmen, farmers and merchants. The merchants traded the produce of the farms for wine, tobacco, wool, skins and fruits.

The Jewish community cared for its sick, its widows and orphans, and supported religious institutions.

In the early part of the twentieth century, a Blood Libel was perpetrated against the Jews of Kishinev. Hooligans swarmed around the Jewish section of Kishinev and proceeded to wipe out much of the Jewish community. Mob violence was responsible for the loss of hundreds of lives, lost businesses and ransacked homes. Terror hung over the Jewish section of Kishinev for two years.

The police contributed to the disaster by standing idly by. They interfered only when Jews attempted to defend themselves. The Tzar pretended to investigate superficially, but in reality, he instigated the continuation of the pogroms. The civilized world protested and passed resolutions of sympathy, but the Tzar refused to listen to demands for justice from his more liberal subjects and from representatives of western democracies.

This story is about a merchant who lived in Kishinev during the Blood Libels of 1903-1905. He escaped with his life but lost his business. He could not extricate himself from a state of despair.

The Jew refused to face the world as a pauper, so he decided to commit suicide. However, he wanted his death to leave a dramatic impact upon the citizens of Kishinev. He went to the pharmacist, purchased a small bottle of poison and took it to the synagogue. Once there, he moved a bench facing the book shelves, climbed onto it and set the bottle of poison securely on the top shelf between two books. He decided that he would wait a few days before he actually took the drastic step of taking his own life.

A few days later, determined to carry out his plan, he returned to the synagogue. As he was about to retrieve the bottle of poison from the top shelf, his hand inadvertently knocked over one of the books. It fell to the floor open. He stooped to see which book was lying on the floor and he noticed that it was Likutay Maharan, the teachings of Rebbe Nachman of Bratslav. He picked up the book and began reading the open page. He read, in amazement, Rebbe Nachman's admonition to the Jewish people:

> "To my fellow Jews, no matter what happens to you, 'zei zich nisht meyaesh', do not despair!"

He read the words over and over. He struggled with the meaning of the words. He wondered how Rebbe Nachman could possibly have known about his horrible plight. He decided to wait one more day to think over his decision to commit suicide.

He returned home and noticed a letter in his pile of mail which was written in a vaguely familiar hand. It brought back memories of an old friendship. He opened the letter and found a check for a sizable sum of money enclosed with a letter. The writer of the letter explained that the check was the repayment of a debt that had been borrowed many years ago.

He recalled that he had never expected repayment on that debt. He had completely forgotten about it. The writer of the letter explained that he had inherited a fortune and was now repaying the old debt.

The despairing man was very surprised at this turn of good fortune. He was so happy that he had struggled with Rebbe Nachman's words of admonition.

He arranged to put his financial affairs in order with the repaid debt. He recouped some of his losses.

In gratitude to Rebbe Nachman, he vowed to reprint and distribute the book which saved his life to Jewish institutions and communities. He wanted to share with as many Jews as possible the words of Rebbe Nachman's admonition:

"Zei zich nisht meyaesh, do not despair!"

1. Rebbe Natan. Atarah L'Rosh Tzaddikim. Collected stories and teachings of Rebbe Nachman of Bratslav, written for posterity by his disciple Rebbe Natan. Reprinted in New York, 1983, by Rabbi Schick, a Bratslaver Chassid.

WE WILL BE HERE TOMORROW[1]

Poland was one of the great centers of Jewish life for centuries, but the perpetrators of the Holocaust annihilated all traces of a once proud and distinguished Polish Jewry. The setting of this story is the Warsaw Ghetto, Poland. It is the last story in this chapter. It sums up in a few words the faith and stubborn will of the Jewish people to live.

Mordechai was sitting with his family around the Sayder table in the Warsaw Ghetto. Houses on the edge of the ghetto burned, bombs exploded all around, and the sound of marching soldiers drowned out the cries of the inhabitants as the resistance fighters bravely tried to withstand the onslaught.

Mordechai turned to his father and said:

> "I have already asked you the four questions concerning the Exodus from Egypt, and now I would like to ask a fifth question. Father, will I be here tomorrow?"

Tears streamed down his father's face as he turned to his young son and answered:

> "My dearest son, I cannot promise that you will be here tomorrow, but I swear to you that there will always be Jewish children in this world who will perpetuate that which is holy to the Jewish people."

It is now forty years after the Holocaust. I would like to reiterate the words which Mordechai's father gave him.

> "Yes, Mordechai, we will be here tomorrow."

1 This is an original story by the author.

BIBLIOGRAPHY

TANACH - TORAH

B'Rayshit (Genesis)
Sh'mot (Exodus)
Vayikra (Leviticus)
Devarim (Deuteronomy)

NEVIIM (THE PROPHETIC WRITINGS)

Yehoshua (Joshua)
Sh'muel (Samuel I and II)
Melochim (Kings I and II)
Yeshayahu (Isaiah)
Yermeyahu (Jeremiah)
Yechezkel (Ezekiel)
Micha (Mica)
Zechariah (Zacharaya)

KETUVIM (HAGIOGRAPHA)

Tehillim (Psalms)
Mishlay (Proverbs)
Iyuv (Job)
Aycha (Lamentations)

TALMUD BAVLI

Order	Tractate
ZERAIM	B'rachot
MOED	Shabbat
	Pesachim
	Yoma
	Succa
	Baytza
	Taanit
	Megilla
	Chagiga
NASHIM	Ketuvot
	Nedarim
	Gitin
NEZIKIN	Baba Metzia
	Baba Batra
	Sanhedrin
	Makkot
	Shevuot
	Edyot
	Avoda Zara
	Avot
KADASHIM	Bechorot

MIDRASH

Aycha Raba
Ayleh Ezkarah
B'Rayshit Raba
Sifra, Kedoshim
Tanna Devay Eliyahu
Vayikra Raba
Yalkut Reuvayni, B'Rayshit

ZOHAR

Zohar, Acharay Mot

ENGLISH SOURCES

BRODY, HEINRICH, ed. *Selected Poems of Yehuda HaLayvi*. The Jewish Publication Society, Philadelphia, 1974.

BUBER, MARTIN. *Legends of the Baal Shem*. Schocken Books, New York, 1969.

FLAVIUS, JOSEPHUS. *Antiquities of the Jews: War of the Jews*. Translated, Whiston. Bigelow, Brown and Company.

IBN DAUD, ABRAHAM BEN DAVID HALAYVI. *The Book of Tradition: An Authoritative History of Spanish Jewry*.

KLASS, RABBI SHOLOM. *Tales of our Gaonim*. Balshon Printing Company, Brooklyn, N.Y., 1967.

KOLITZ, ZVI. *The Tiger Beneath the Skin*. *"Yossel Rakover's Appeal to G-d"*. Farrar, Strauss and Giroux, Inc., New York.

KRAVITZ, NATHANIEL. *Three Thousand Years of Hebrew Literature*. The Swallow Press, Chicago, 1972.

LEHMAN, MARCUS. *Five Novellettes*. Merkos L'Inyonay Chinuch, Inc., Brooklyn, N.Y., 1964.

MEISELS, MEIR. *Judaism*. Feldheim Publishers, Jerusalem.

ROSMARIN, TRUDE WEISS. *Jewish Survival*. Philosophical Library, New York, 1949.

URY, RABBI ZALMAN. *The Story of Rabbi Yisrael Salanter*. Torah U'Mesorah Publications, New York, 1971.

GERMAN SOURCES

Sippurim Prager Sammlung: Judisher Legenden. R. Lowit Verlag, Wien und Leipzig, 1921.

HEBREW SOURCES

BARUCH, Y.L. *Sayfer Hamoadim: Rosh Hashana V'Yom Kippur.* (*Vol. I*). The Oneg Shabbat (Ohel Shem) Society. Dvir Publishing Co., Tel Aviv, 1952.

BEKMEISTER, YISRAEL. *Sipuray Niflaot Migdolay Yisrael.* Ayshel Press, Benay B'rak, 1967.

Chumash Bayt Yehudah Hashalem. Hotzaat Orah, New York, 5714 (1954).

DICK, I.M. *Garay HaTzedek.* Vilna, 1862.

EISENSTEIN, J.D., ed. *Otzar Yisrael.* Shiloh Publishing Co., Jerusalem.

ELIAV, MORDECHAI. *Ani Maamin.* Mosad Harav Kook, Jerusalem, 1969.

GETZ, MENACHEM. *Yerushalayim Shel Maaleh.* Haginzar Hacharaydi, Jerusalem, 1977.

JELLINEK, AARON. *Bayt Hamidrash.* Bamberger and Wahrman, Jerusalem, 1938.

KASHER, M.S. and Y.Y. BELKROVITZ. *Anshay Maaleh.* Maarav Publishing Co., Jerusalem, 1974.

KASHER, M.S. and Y.Y. BELKROVITZ. *Anshay Mofayt.* M.S.D. Publishing Co., Jerusalem. 1976.

KASHER, M.S. *Baalay Teshuva.* Maarav Publishing Co., Jerusalem, 1974.

KI TOV, ELIYAHU. *Sayfer Haparsheyot: Kedoshim.* Alef Publications, Jerusalem, 1970.

KI TOV, ELIYAHU. *Sayfer Hatodaah.* Yeshurun Publishers, Jerusalem, 1966.

KATZ, YAAKOV AND HERSHKO, MOSHE. *Yisrael V'Haamim.* Dvir Publishing Co., Tel Aviv, 1961.

KOMARNA, RABBI YITZCHAK. *Haychal Haberacha.*

LIPKIN, CHAIM YITZCHAK. *Taryag Mitzvot B'Halacha U'V'Agada.* Netzach Publishers, Benay B'rak, 1962.

MAIMONIDES. *Matnat Evyonim.*

MAIMONIDES. *Shlosha Asar Ikarim.*

REBBE NATAN. *Atara L'Rosh Tzaddikim.* Bratslav.

PRAGER, MOSHE. *Nitzotzay Gevura.* Jewish Education Committee of New York, New York, 1966.

REBBE CHAIM ELEAZAR SHAPIRA. *Sayder Shana Haacharona.* Munkatch.

Shivchay Tzaddikim. Bayt Hillel Publishing House, New York, 1980.

Sh'moneh Seforim: Tzeror Hachayim and Bayt Pinchas. Printing House of Rabbi Natan Neta Kronenberg, Bulgaria, 1903.

TSCHERNIKOVSKY, SAUL. *Mivchar Shaul Tschernikovsky.* Dvir Publishing Company, Tel Aviv, 1964.

REBBE SHNEOR ZALMAN OF LIADI. *Tanya.* Slavita, 1796.